T0299252

Theoretical Insights into the Electrochemical Properties of Ionic Liquid Electrolytes in Lithium-Ion Batteries

This book provides a concise overview of the use of ionic liquids as electrolytes in lithium-ion batteries (LIBs) from a theoretical and computational perspective. It focuses on computational studies to understand the behavior of lithium ions in different ionic liquids and to optimize the performance of ionic liquid-based electrolytes. The main features of the book are as follows:

- Provides a thorough understanding of the theoretical and computational aspects of using ionic liquids as electrolytes in LIBs, including the evaluation and reproducibility of the theoretical paths.
- Covers various computational methods such as density functional theory, molecular dynamics, and quantum mechanics that have been used to study the behavior of lithium ions in different solvents and to optimize the performance of ionic liquid-based electrolytes.
- Discusses recent advances such as new computational methods for predicting the properties of ionic liquid-based electrolytes, new strategies for improving the stability and conductivity of these electrolytes, and new approaches for understanding the kinetics and thermodynamics of redox reactions with ionic liquids.
- Suggests how theoretical insights can be translated into practical applications for improving performance and safety.

This monograph will be of interest to engineers working on LIB optimization.

Leila Maftoon-Azad is Assistant Professor, College of Nano and Bio Sciences and Technologies, Persian Gulf University, Iran. She specializes in computational chemistry with a focus on the development of energy materials for applications in lithium-ion batteries, adsorbents, and solar cells.

Theoretical Insights into the Electrochemical Properties of Ionic Liquid Electrolytes in Lithium-Ion Batteries

Leila Maftoon-Azad

CRC Press
Taylor & Francis Group
Boca Raton London New York

CRC Press is an imprint of the
Taylor & Francis Group, an **informa** business

First edition published 2025
by CRC Press
2385 NW Executive Center Drive, Suite 320, Boca Raton FL 33431

and by CRC Press
4 Park Square, Milton Park, Abingdon, Oxon, OX14 4RN

CRC Press is an imprint of Taylor & Francis Group, LLC

© 2025 Leila Maftoon-Azad

Library of Congress Cataloging-in-Publication Data
Names: Maftoon-Azad, Leila, author.
Title: Theoretical insights into the electrochemical properties of ionic liquid electrolytes in lithium-ion batteries / Leila Maftoon-Azad.
Description: First edition. | Boca Raton, FL : CRC Press, 2025. | Includes bibliographical references and index.
Identifiers: LCCN 2024020824 (print) | LCCN 2024020825 (ebook) |
ISBN 9781032866031 (hardback) | ISBN 9781032872780 (paperback) |
ISBN 9781003531821 (ebook)
Subjects: LCSH: Lithium ion batteries. | Electrochemistry.
Classification: LCC TK2945.L58 M34 2025 (print) | LCC TK2945.L58 (ebook) |
DDC 621.31/2424—dc23/eng/20240806
LC record available at https://lccn.loc.gov/2024020824
LC ebook record available at https://lccn.loc.gov/2024020825

ISBN: 978-1-032-86603-1 (hbk)
ISBN: 978-1-032-87278-0 (pbk)
ISBN: 978-1-003-53182-1 (ebk)

DOI: 10.1201/9781003531821

Typeset in Times LT Std
by codeMantra

Contents

Preface

In the name of God, the merciful, the compassionate.

Lithium-ion batteries (LIBs) have become ubiquitous in portable electronics, electric vehicles (EVs), and grid-scale energy storage systems due to their high energy density, long cycle life, and low self-discharge rate. While they have revolutionized technology and facilitated the widespread adoption of EVs and renewable energy sources, they are not without limitations. Safety concerns, limited energy density, and challenges related to the use of ionic liquid (IL) electrolytes have prompted the need for further research and development in this field.

In response to these limitations, researchers have turned to computational methods to gain a deeper understanding of the fundamental processes that govern the behavior of LIBs. This has led to significant advances in the field, enabling the accurate prediction of key properties and the identification of potential solutions to address existing challenges. However, the use of IL electrolytes in LIBs presents its own set of challenges, including safety concerns, high cost, and limited scalability.

To address these issues, researchers have explored various theoretical approaches, such as modifications to the chemical structure of ILs and their combinations with other materials. While these efforts have shown promise, there is a need to thoroughly evaluate the validity, data management, and reproducibility of the theoretical approaches used in these studies. By establishing a set of guidelines for validating, managing, and reproducing theoretical approaches, researchers can contribute to a more robust and reliable field of LIB research.

The potential of LIBs as an efficient energy storage system is undeniable, and ongoing research and enhancement strategies are essential to further improve their energy and power density. By addressing the current limitations and challenges, researchers can pave the way for the continued advancement and widespread adoption of LIBs in various energy storage and conversion applications. In this work, we provide an overview of the issues that have been associated with the use of ILs in LIBs. The safety concerns, high cost, and limited scalability of ILs have been identified as significant issues that need to be addressed in order to fully realize the potential of ILs in LIBs. To address these challenges, various theoretical approaches have been explored in the literature, including modifications to the chemical structure of ILs and combinations with other materials.

Through a comprehensive review of the literature, we have identified several studies that have investigated the challenges associated with the use of ILs in LIBs. However, the validity, data management, and reproducibility of theoretical approaches used in these studies have not been thoroughly evaluated. To address these issues, we have conducted a review of the literature and identified several key factors that contribute to the validity, data management, and reproducibility of theoretical approaches. These factors include the use of appropriate models, the rigor of data collection and analysis, and the transparency of methodology and results.

To provide a framework for addressing these issues, we propose a set of guidelines for validating, managing, and reproducing theoretical approaches used in LIB research. These guidelines emphasize the importance of using appropriate models, conducting rigorous data collection and analysis, and ensuring transparency in methodology and results. By following these guidelines, researchers can improve the validity, data management, and reproducibility of theoretical approaches and contribute to a more robust and reliable field of LIB research.

In crafting this book, I have harnessed the power of artificial intelligence (AI) to explore diverse realms of knowledge. By leveraging AI-generated questions and refining the answers through extensive research in academic journals and books, I have delved into new frontiers. The process involved feeding the responses to another AI system and further enhancing and polishing them under human oversight and direction. While AI technology has been instrumental in shaping the content of this book, the final work is the result of careful curation and refinement under human leadership.

Introduction

1

Lithium-ion batteries (LIBs) are widely used in portable electronics, electric vehicles (EVs), and grid-scale energy storage systems (1–4). This is attributed to their high energy density, long cycle life, and low self-discharge rate (5–10). They have revolutionized the way we use technology and have enabled the widespread adoption of EVs and renewable energy sources (11–14).

However, LIBs also have some limitations. One major limitation is their safety concerns, as they have a tendency to overheat and become unstable and can catch fire or explode if not handled properly (15–19). Another limitation is their limited energy density, which limits the range of EVs (20,21) and the amount of energy that can be stored in grid-scale energy storage systems (1,22,23). The driving range of EVs is a primary concern for customers and is determined by the energy density of the batteries. LIBs have high energy storage densities but fall short of gasoline. Next-generation batteries with energy densities beyond LIBs are needed to increase driving range effectively. New designs such as Li-sulfur, Li-air, or Mg-ion batteries have higher theoretical energy densities but suffer from safety or poor recyclability issues. Battery packs in EVs also include other components that reduce overall energy densities. Improving cell design and pack efficiency is critical to increasing the energy densities of EV batteries (20).

Ionic liquid (IL) electrolytes have been proposed as a potential solution to these limitations. ILs are salts liquid at or near room temperature and have unique properties that make them attractive as electrolytes in batteries. However, using ILs in lithium batteries also poses several challenges, including safety concerns, high cost, and limited scalability.

In this work, we provide an overview of the issues that have been associated with the use of ILs in LIBs. The safety concerns, high cost, and limited scalability of ILs have been identified as significant issues that need to be addressed in order to fully realize the potential of ILs in LIBs. To address these challenges, various theoretical approaches have been explored in the literature, including modifications to the chemical structure of ILs and combinations with other materials.

DOI: 10.1201/9781003531821-1

This book aims to provide a comprehensive review of the theoretical approaches used in the study of LIBs that utilize ILs as electrolytes. Despite the growing interest in the use of ILs in LIBs, the validity, data management, and reproducibility of the theoretical approaches used in this field have not been thoroughly evaluated. To address this gap, this book identifies several key factors that contribute to the reliability and robustness of theoretical approaches in LIB research, including the use of appropriate models, rigorous data collection and analysis, and transparency in methodology and results. The book proposes a set of guidelines for validating, managing, and reproducing theoretical approaches used in LIB research, with an emphasis on the importance of using appropriate models, conducting rigorous data collection and analysis, and ensuring transparency in methodology and results. The book also provides a critical review of the literature on the use of ILs in LIBs, highlighting the challenges associated with ILs in terms of ion transfer, both in bulk and at the electrode interface. The book presents an overview of the computational methods used to study the case, including their evaluation and data management, with a focus on proper reproducibility.

Ionic Liquids, Advantages and Limitations

2

Ionic liquids (ILs) have several properties and characteristics that make them suitable for use as electrolytes in lithium batteries.

One of the critical advantages of ILs is their high thermal stability. They have very low vapor pressure and do not evaporate easily, which makes them less prone to combustion or explosion than some other types of electrolytes. This fact is particularly important for high-energy-density batteries like lithium-ion batteries, which can generate much heat during charging and discharging.

ILs also have low flammability and toxicity, which makes them safer to handle and dispose of than some other types of electrolytes. They are also non-volatile, which means that they do not produce gas during cycling, reducing the risk of gas build up and potential rupture of the battery (24–29).

Another advantage of ILs is their wide electrochemical window. It means they can withstand a wide range of voltage potentials without breaking down or decomposing. This is important for lithium batteries because the voltage range in these batteries can be pretty high, and the electrolyte needs to withstand this without reacting or degrading. Thus, ILs can potentially increase the energy densities of lithium batteries (30–35).

Finally, ILs have a high degree of tunability. This implies that the characteristics of these materials can be customized to suit specific purposes. This allows for developing customized electrolytes with optimal properties for a particular battery design.

Overall, the combination of high thermal stability, low volatility, wide electrochemical window, and tunability make ILs a promising choice for electrolytes in lithium batteries. These properties and characteristics enable the development of safer, more efficient, and longer-lasting batteries.

DOI: 10.1201/9781003531821-2

However, there are still challenges that need to be addressed, such as their high viscosity, low ionic conductivity, high cost, and limited availability compared to traditional electrolytes (36–41).

ILs can have a relatively high viscosity, which can limit their ionic conductivity and diffusion rates in the battery. This can lead to performance issues such as low power density and poor cycling stability (42–45).

ILs can be relatively expensive to produce, making them less competitive with other types of electrolytes in terms of cost. The primary reason for this is the utilization of various chemical substances during the synthesis procedure, coupled with advanced purification methods (46–49).

While ILs are generally stable over a wide range of voltages and temperatures, some types may be less stable under certain conditions. For example, some ILs can decompose or react with the lithium electrode at high potentials or temperatures, reducing battery performance or posing safety risks (50).

ILs may have limited solubility for some types of lithium salts, which can affect their ionic conductivity and overall battery performance (51,52).

Integrating ILs into existing battery designs may be challenging due to their unique properties and requirements. This may require modifications to the battery design or the development of new manufacturing processes (53–58).

Addressing these challenges will require continued research and development to optimize the properties of ILs as electrolytes for lithium batteries. This may involve the development of new types of ILs with lower viscosity, higher solubility, and improved stability, as well as the optimization of battery design and manufacturing processes to integrate ILs more effectively. Additionally, cost-reduction efforts will be necessary to make ILs more competitive with other electrolytes. Despite these challenges, the potential benefits of IL electrolytes, such as improved safety, higher energy density, and longer cycle life, make them an area of active research and development in the field of advanced battery technologies (59).

STRATEGIES FOR OVERCOMING LIMITATIONS OF IL ELECTROLYTES

There are several ways to address the limitations of IL electrolytes.

Viscosity

To overcome the issue of high viscosity, researchers are exploring ways to modify the structure of ILs or add small amounts of other solvents to reduce their viscosity. For example, researchers are exploring the utilization of functionalized ILs or the addition of small amounts of co-solvents to improve the ionic conductivity and reduce the viscosity of the electrolytes.

Modifying the Structure of Ionic Liquids to Reduce Viscosity

Modifying the structure of ILs is one way to reduce their viscosity. One approach is to introduce structural features that disrupt the packing of ions and reduce the strength of interionic interactions. This can be achieved by varying the size, shape, and, or functional groups of the ions that make up the IL. For example, introducing alkyl chains of different lengths or branching patterns can increase the distance between the ions and reduce the strength of interionic interactions, leading to lower viscosities.

Another approach is introducing functional groups that can weaken the interionic interactions and reduce viscosity.

Tansel (60) focused on the importance of thermodynamic and physical characteristics in the permeation of ions during membrane separation. Specifically, the hydrated radius, hydration-free energy, and viscous effects were studied concerning how they impact ion transport through a membrane. The authors suggested that a deeper understanding of these factors can help to optimize membrane separation processes and improve the efficiency of ion transport.

Izgorodina et al. (61) explored the role of dispersion forces in predicting the thermodynamic and transport properties of common ILs. The authors suggested that dispersion forces, a type of intermolecular force, play a crucial role in determining the ILs properties, including viscosity, density, and thermal conductivity. The authors use an assembly of experimental measurements and molecular simulations to investigate the influence of dispersion forces on these properties. They concluded that accurate modeling of dispersion forces is essential for predicting the ILs characteristics and designing new ILs with tailored properties.

Dean et al. (62) discussed the importance of structural analysis in understanding the behavior of low-melting organic salts, particularly in the context of ILs. The authors reviewed various techniques that can be used to study the structure of these materials, including X-ray crystallography and NMR spectroscopy. They also discussed the importance of understanding the structural properties of ILs, such as their intermolecular interactions and the organization of their constituent ions, in relation to their unique physical and chemical properties.

Sanchora et al. (63) highlighted the importance of considering the effects of the alkyl chain length and water content on the properties of ILs. Using a combination of experimental and computational techniques, the authors explored how changes in the alkyl chain length and water content affect the 1-alkyl-3-methylimidazolium chloride IL's physical and chemical properties, such as its ion pairing energy and hydrogen bonding which impact density, viscosity, and surface tension.

Molecular dynamic simulations provide valuable insights into the reason behind the lower viscosity observed when alkoxy chains are incorporated instead of alkyl chains. These simulations suggest that the reduced viscosity result from the alkoxy chains being less efficient in assembling and interacting with each other, leading to minimal aggregation. Experimental findings support this, showing that the presence of alkoxy chains reduces intermolecular correlations and cation-anion electrostatic interactions, resulting in faster dynamics compared to alkyl counterparts. Raman-induced Kerr spectroscopy studies demonstrate that including ether-substituted groups weakens interionic interactions due to their larger volume. However, their flexibility allows for faster reorientation and stronger interionic interactions. Moreover, certain imidazolium and pyridinium cations produce stable complexes with polyethylene glycol (PEG) chains through ion-dipole interactions. Consequently, ILs containing long and flexible alkoxy chains are expected to exhibit diminished Coulombic interactions between the cation and anion species (64–69).

Fumino et al. (70) utilized far infrared and terahertz spectroscopy to study the interactions in Coulomb fluids. The results indicated that the overall interaction between cations and anions in ILs is a delicate balance between Coulomb forces, hydrogen bonding, and dispersion forces. In the case of protic ILs, the low-frequency spectra showed distinct vibrational modes that revealed the presence of medium to strong hydrogen bonds between the cations and anions. The researchers also employed isotopic substitution to isolate frequency shifts related to interaction strength and reduced masses.

The study further investigated how these interactions impact the physical properties of ILs, such as their melting point, viscosity, and enthalpy of vaporization.

Hayyan et al. (71) conducted a study to synthesize Deep Eutectic Solvents (DESs) by combining triethylene glycol (TEG) with five different phosphonium and ammonium salts. They examined the physical properties of these synthesized DESs at different temperatures ranging from 25°C to 80°C. Fourier transform infrared spectroscopy (FTIR) was also used to analyze the functional groups present in the DESs. The physical properties of deep eutectic solvents (DES) were significantly influenced by blending either ammonium or phosphonium salts with triethylene glycol (TEG) as the hydrogen bond donor (HBD). These properties include freezing point, viscosity, electrical conductivity, and density. The study demonstrated that the physical properties of DESs can be tailored by selecting appropriate HBDs and salts.

Briefly, modifying the structure of ILs to reduce their viscosity involves balancing the competing effects of interionic interactions and structural features that can disrupt these interactions. By carefully designing the structure of ILs, it is possible to achieve the desired balance and create ILs with lower viscosities for specific applications.

Co-solvent Addition

The choice of co-solvent depends on the specific application and performance requirements, as well as the compatibility of the co-solvent with the IL and the electrode materials. It is essential to carefully select co-solvents to ensure that they do not adversely affect the stability and safety of the battery. In addition, the amount and type of co-solvent added should be optimized to balance the trade-off between viscosity reduction and ionic conductivity improvement. Some co-solvents that are appropriate to be added to ILs to reduce viscosity are introduced below:

- *Organic Solvents*: Organic solvents such as acetonitrile, propylene carbonate, and dimethyl carbonate can be added to ILs to reduce their viscosity and improve their ionic conductivity. These solvents can also improve the solubility of lithium salts in ILs, which can further improve their electrochemical performance (72–78).
- *Water*: Adding small amounts of water to ILs can reduce their viscosity and improve their ionic conductivity (79–83). Water can also improve the solubility of lithium salts in ILs and promote the formation of stable solid electrolyte interphase (SEI) layers on electrode surfaces (84).
- *Ionic Liquids*: Mixing two or more ILs with different molecular structures and properties can result in a decrease in viscosity and an increase in ionic conductivity (85–88). This is because the different ILs can interact with each other to form a more fluid and mobile mixture.

  For instance, in the mixing of PAN/BAN and [C2MIM][BF4] ILs, the [C2MIM] cations integrate into protic networks, while the [BF4] anions occupy previously vacant regions near protic cation tails. This subtle microstructural adjustment results in complex variations in the transport properties of the ions. Similarly, for EAN–[C2MIM][BF4] mixtures, a novel conductivity curve exhibits pronounced deviations from the simple ideal mixing rule, with three different regions defined by a local maximum and a global minimum at intermediate concentrations. These regions are defined by the onset of the formation of EAN HB networks and the virtual disappearance of aprotic IL structures, where long-range ordering for [C2MIM][BF4] breaks down (89).

- *Surfactants*: Surfactants can be added to ILs to reduce their surface tension and improve their wetting properties (90). This can improve the contact between the electrolyte and electrode surfaces, improving battery performance. Contact angle testing and electrolyte absorption are commonly used to measure wettability, with a lower contact angle indicating better wettability.

Co-solvent Considerations

When selecting co-solvents for ILs, there are several safety concerns that need to be considered. Some of these concerns include:

- *Volatility*: Some co-solvents can be highly volatile, which can increase the risk of flammability and explosion (91–93). When selecting co-solvents, it is crucial to choose ones that have low volatility and are stable under the conditions of battery operation.
- *Toxicity*: When selecting co-solvents, it is essential to choose ones that are non-toxic and have a low environmental impact (94,95). To assess the environmental impact of co-solvents, it is essential to consider their life cycle, from manufacturing to disposal, through a life-cycle assessment (LCA). However, this data is often tailored to specific applications and is not always available. In practice, it is more realistic to assess solvents on key properties for which data is available, such as hazard labels, physical properties, or biobased feedstock percentage.
- *Compatibility*: When selecting co-solvents, it is vital to ensure that they are compatible with the chosen electrode materials and other components of the battery (96,97). This can be achieved by considering the co-solvent's impact on the battery's ionic conductivity, electrode compatibility, and overall safety during the selection process. Novel cosolvent mixtures have been developed for cutting-edge uses like rapid charging or suitability with lithium-metal electrodes. These mixtures were created using similar design strategies, considering the compatibility of the cosolvent with the battery's electrode materials and other components (98).
- *Stability*: Some co-solvents may not be stable under the conditions of battery operation, leading to degradation and reduced battery performance. It is important to consider the battery's operating

conditions, such as temperature, pressure, and the presence of other chemicals. The co-solvent should also not react with the electrolyte or other battery components, as this could cause degradation and reduced performance. When selecting co-solvents, it is important to choose ones that are stable under the conditions of battery operation.

- *Cost*: Some co-solvents such as fluorinated ones may be more expensive than others, which can increase the overall cost of battery production (99). When selecting co-solvents, it is important to balance the cost with the desired performance and safety requirements.

To address these concerns, it is important to carefully evaluate the properties of co-solvents and their potential impact on battery performance and safety. This can be done through experimental and computational methods, such as electrochemical measurements, spectroscopy, and molecular simulations.

In addition, it is essential to follow best practices for handling and disposing of co-solvents to minimize their impact on human health and the environment. This includes using appropriate personal protective equipment, ensuring proper ventilation, and properly storing and disposing of co-solvents according to local regulations (100).

Cost Limitations

To address the cost issue, researchers are investigating alternative methods of IL synthesis that are more cost-effective. For example, some researchers are exploring the use of renewable or waste materials as starting materials for IL production or recovery (101–103) or lithium battery waste (104). Additionally, the use of ILs in high-value applications, such as aerospace and defence may help to justify their higher cost (105–108).

There are primarily two main methods for the preparation of ILs: metathesis of a halide salt with a desired anion and acid-base neutralization reactions. These methods typically require the use of molecular solvents, which can be expensive and contribute to the overall cost of IL production (109). However, recent research has focused on developing novel methods of synthesis that replace molecular solvents with ILs themselves, which can be more cost-effective. Although ILs are typically considered expensive compared to traditional solvents, their ease of recycling makes them a favorable option for various applications. Researchers are exploring techniques such as membrane separation, extraction, and distillation to recover ILs. These methods have the potential to lower the overall cost of IL production and enhance their sustainability (110).

Stability

To improve the stability of ILs, researchers are exploring new types of ILs that are more stable under high potentials or temperatures. These new ILs are designed to have higher thermal stability, lower polarization, and lower loss of active material at elevated temperatures. The anions in ILs have a significant influence on their tribological properties, with hydrophobic anions such as BF4 and PF6 may cause corrosion of steel under humid conditions. However, other hydrophobic anions such as bis(fluorosulfonyl)imide (FSI) anion are less corrosive and exhibit good tribological properties (111). Additionally, the use of additives or coatings on the electrode or separator (112–116) can help to minimize the interaction between the IL and other materials in the battery.

Solubility

To address the issue of limited solubility, researchers are exploring using different types of lithium salts or modifying the structure of the IL to improve its solubility. For example, some researchers are exploring the use of functionalized ILs or the addition of small amounts of co-solvents to improve the solubility of the electrolytes, which has been discussed in the viscosity section.

Toxicity

To address the issue of toxicity, researchers are exploring the use of ILs with lower toxicity or modifying the structure of the IL to reduce its toxicity. By changing the chemical compositions of the cations and anions, like utilizing non-aromatic compounds, pairing them with particular anions, and adjusting the length and hydrophobicity of the side chains, it is possible to reduce toxicity levels in ILs, rendering them safer for diverse uses. For example, cholinium-based ILs are recognized as the least toxic, while enhancing the hydrophobicity and length of the side chains can elevate toxicity. Hence, the careful selection of cations and anions and the modification of their characteristics are vital in creating less toxic ILs. Additionally, proper handling and disposal protocols can help to minimize the risk of exposure to toxic ILs (117–122).

Integration

To address the issue of integration, researchers are exploring the use of new battery designs or modification of existing designs to better accommodate the unique properties of IL electrolytes. This may involve the development of new manufacturing processes or the use of specialized equipment to handle and store the electrolytes (123–127). The integration of IL electrolytes in battery systems requires careful consideration of the electrolyte selection, design, and manufacturing processes. The use of simulation solutions, additives or coatings, and the development of new manufacturing processes or specialized equipment can help optimize battery design and engineering, making them more stable, safer, and cost-effective. Simulation solutions for all physics (chemical, electrical, mechanical, thermal) and scales (from material to cell, module, pack, full vehicle integration) are also being developed to optimize battery design and engineering.

Overall, addressing the limitations of IL electrolytes will require continued further research and innovation to enhance their functionality and performance for specific battery applications. This may involve a combination of modifying the structure of the IL, developing new synthesis methods, and optimizing battery designs and manufacturing processes.

Some New Battery Designs that can Better Accommodate the Unique Properties of Ionic Liquid Electrolytes

There are several new battery designs that can better accommodate the unique properties of IL electrolytes. Some examples include the following:

- *Solid-State Batteries*: Solid-state batteries use a solid electrolyte instead of a liquid electrolyte, which can improve safety, stability, and energy density, Figure 1. IL electrolytes can be used as solid electrolytes or as additives to improve the properties of the solid electrolyte. For example, ILs can improve ionic conductivity and reduce the interfacial resistance between the solid electrolyte and the electrodes (128–135).
- *Flow Batteries*: Flow batteries use a liquid electrolyte stored in external tanks and circulated through the battery during operation. IL electrolytes can be used as the electrolyte in flow batteries, which can improve energy density and reduce the risk of leakage or combustion (136–143).

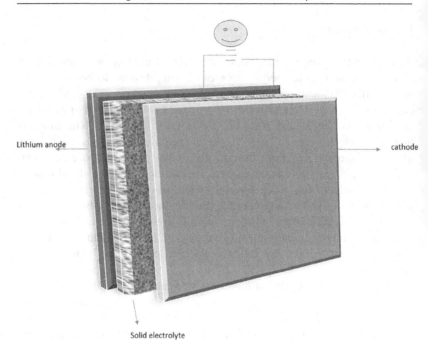

FIGURE 1   Schematics for a solid state battery.

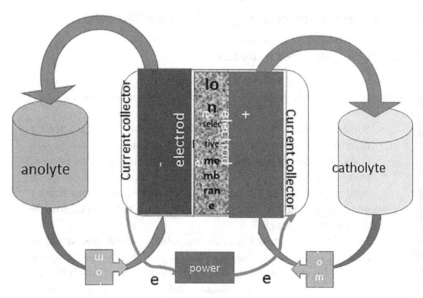

FIGURE 2   Schematics for a typical flow battery.

- *Lithium-Sulfur Batteries*: Lithium-sulfur batteries use a sulfur cathode and a lithium anode, which can provide high energy density and low cost. However, sulfur cathodes typically have poor stability and low conductivity (144–146). IL electrolytes can be used as additives to improve the stability and conductivity of the sulfur cathode, leading to better battery performance (147–153).
- *Three-Dimensional Batteries*: Three-dimensional (3D) batteries use a porous electrode structure that allows for better electrolyte penetration and ion transport (154–157). IL electrolytes can be used in 3D batteries to improve the transport properties of the electrolyte and reduce the concentration polarization at the electrode/electrolyte interface, leading to better battery performance (158,159).
- *Electrolyte-Filled Batteries*: Electrolyte-filled batteries use a porous electrolyte-filled structure instead of a traditional separator. This structure allows for better ion transport and reduces the risk of short circuits or dendrite formation (160–165). IL electrolytes can be used in electrolyte-filled batteries to improve their stability and reduce the risk of leakage or combustion (166–170).

Overall, these new battery designs can better accommodate the unique properties of IL electrolytes and improve their performance in terms of energy density, safety, and stability. However, developing these new battery designs will require significant research and development to optimize their properties and performance. Additionally, the high cost of IL electrolytes may be a limiting factor in some of these designs, and further cost-reduction efforts will be necessary to make them more competitive with other types of electrolytes.

Ion Transport, Electrode-Electrolyte Interface

3

The ion transport mechanism in ionic liquid electrolytes differs from traditional organic solvent-based electrolytes. In ionic liquids, ions can move through the bulk liquid or along the surface of the electrode, depending on the specific properties of the electrolyte and the electrode. This can impact the battery's performance, as different ion transport mechanisms may have different rates of ion transfer and different levels of resistance.

The impact of these properties on the performance of the lithium battery can vary depending on the specific application and battery design. For example, in high-power applications, such as electric vehicles, high ionic conductivity and fast ion transport are important for achieving high power output. However, in high-energy-density applications, such as stationary energy storage, the stability of the electrolyte is more important, as the battery must be able to operate reliably over many cycles without degradation.

The viscosity of the ionic liquid electrolyte can impact the rate of ion transport and diffusion in the battery. Higher viscosity can lead to slower ion transport, resulting in lower power output and reduced cycling stability. To address this issue, researchers are exploring ways to modify the structure of the ionic liquid or add small amounts of co-solvents to reduce the viscosity of the electrolyte.

The ion transport mechanisms in ionic liquid electrolytes can also affect the battery's performance. For example, if the ion transport occurs predominantly through the surface of the electrode, the battery may experience higher resistance and lower capacity.

The electrode-ionic liquid interface is an essential component of the lithium battery, as it governs the transfer of ions between the electrode and the electrolyte. Here are a few ways that researchers are working to optimize the electrode-ionic liquid interface to improve ion transport and reduce resistance:

                    DOI: 10.1201/9781003531821-3

A. *Surface Modification*: One approach to optimizing the electrode-ionic liquid interface is to modify the surface of the electrode to improve its wettability and reduce the interfacial resistance (171–173). For example, researchers have experimented with applying a thin layer of metal oxide or a conductive polymer coating to the surface of the electrode to improve its interaction with the electrolyte (174–176). Researchers have also used ionic liquid as a wetting agent for the interface between solid-state electrolytes and electrodes to enhance interfacial wetting and improve battery performance. The optimization of the ionic liquid content in the interface has been crucial for achieving high performance in solid-state batteries (177).

B. *Nanostructuring*: Another approach is to modify the surface of the electrode at the nanoscale. Nanostructuring can increase the surface area of the electrode and improve its interaction with the electrolyte (178–183). Several materials are considered promising for nanostructured electrodes in lithium-ion batteries. These materials possess unique properties that make them attractive for electrode applications, such as high surface area, high conductivity, and good stability. Here are a few examples:

a *Carbon Nanotubes (CNTs)*: CNTs are one-dimensional structures made of carbon atoms arranged in a cylindrical shape. They have a high surface area and excellent electrical conductivity, making them attractive for use as electrode materials in lithium-ion batteries. In addition, CNTs have good mechanical strength and flexibility, which can improve the electrode's durability (184–186).

b *Graphene*: It is a two-dimensional material made of a single layer of carbon atoms arranged in a hexagonal lattice. It has a high surface area, excellent electrical conductivity, and good mechanical properties, making it attractive for use as an electrode material in lithium-ion batteries. Graphene can also be easily functionalized with other materials to improve its performance (187).

c *Metal Oxides*: Metal oxides, such as titanium dioxide (TiO_2) and iron oxide (Fe_2O_3), have attracted attention as potential electrode materials due to their high stability and low toxicity (188,189). Metal oxides can also have a high theoretical capacity, which can further improve the energy density of the battery (190–192). However, their low electrical conductivity can limit their performance as electrode materials (193–195).

d   *Metal Sulfides*: Metal sulfides, such as molybdenum disulfide (MoS_2) and cobalt sulfide (CoS), have also been studied as potential electrode materials in lithium-ion batteries (196–200). Metal sulfides have a high theoretical capacity and good electrochemical stability, making them attractive for use in high-performance batteries (201,202). However, their low electrical conductivity can limit their performance, and more research is needed to improve their conductivity.

e   *Silicon*: It is a promising material for use as an electrode material in lithium-ion batteries due to its high theoretical capacity and abundance (203,204). However, silicon electrodes can suffer from significant volume changes during cycling, which can lead to mechanical degradation and reduced performance. Researchers are exploring ways to mitigate this issue by using nanostructured silicon electrodes, which can improve the mechanical stability of the electrode (205,206).

The smaller size and nanostructured design of these silicon electrodes can help accommodate the volume changes better and prevent cracking, pulverization, and loss of electrical contact that can occur in larger silicon electrodes (207).

However, the commercialization of these delicate nanostructured silicon electrodes is still challenging due to issues like poor first-cycle coulombic efficiency (208) and higher manufacturing costs compared to larger silicon particles.

C. *Electrolyte Additives*: Adding electrolyte additives can also improve the electrode-ionic liquid interface (209–213). For example, researchers have added small amounts of lithium salts or other additives to the electrolyte to improve the wetting of the electrode surface and reduce the interfacial resistance (210). Tuning the reaction time and/or electrolyte composition (e.g., using different lithium salts like LiFSI, LiPF6, and LiAsF6) can lead to diverse surface morphologies on the lithium metal electrode, which can impact the wettability and interfacial resistance (214,215).

D. *Interface Modelling*: Finally, researchers are using computational methods to model and optimize the electrode-ionic liquid interface at the molecular level (216–222). These models can provide insights into the specific interactions between the electrode and the electrolyte, allowing for the design of more effective electrode materials and electrolytes. While computational modelling is a powerful tool for optimizing the electrode-ionic liquid interface in lithium-ion batteries, it has some limitations. Here are a few examples:

a. *Complexity*: The electrode-ionic liquid interface is a complex system with many interacting components, including the electrode material, the ionic liquid electrolyte, and the interface between them. Modeling this system accurately requires a high level of computational complexity and can be computationally expensive, especially for large-scale systems.

b. *Accuracy*: The accuracy of computational models for the electrode-ionic liquid interface depends on the accuracy of the input parameters and assumptions used. For example, the properties of the ionic liquid electrolyte, such as its viscosity and dielectric constant, can vary depending on the specific formulation and conditions, and accurate modelling of these properties can be challenging (223).

c. *Validation*: Validation of computational models for the electrode-ionic liquid interface can be complex, as experimental measurements of the interface are often limited and can be affected by surface roughness, impurities, and sample preparation. As a result, it can be challenging to validate the accuracy of computational models and to ensure that they are representative of the real-world system (224).

d. *Limitations of Current Models*: Current computational models for the electrode-ionic liquid interface often rely on simplifying assumptions and approximations, such as treating the ionic liquid as a continuum medium and ignoring the effects of solvent molecules and other small ions. These approximations can limit the accuracy and applicability of the models, and more advanced models that incorporate more detailed information about the ionic liquid electrolyte are needed (225,226).

e. *Scaling*: Finally, computational modelling is often limited by scalability, as simulating larger systems requires higher computational resources. This can limit the ability of researchers to model the electrode-ionic liquid interface in complex systems, such as in multi-layered electrodes or systems with multiple interfaces (227). Despite these limitations, computational modelling remains an important tool for optimizing the electrode-ionic liquid interface in lithium-ion batteries.

By combining computational modelling with experimental measurements and other analytical techniques, researchers can gain a better understanding of the complex interactions at the interface and develop more effective strategies for improving the performance and stability of lithium-ion batteries (228–232).

Theoretical and Experimental Investigation of Ionic Liquids in Lithium Ion Batteries

4

The behavior of IL electrolytes in lithium batteries has been extensively studied using theoretical models that take into account the thermodynamics and kinetics of the electrochemical reactions. These models play a crucial role in understanding the complex electrochemical processes occurring within batteries and are essential for predicting and optimizing battery performance.

- *Poisson-Nernst-Planck (PNP) Equation*: One theoretical model used to describe the behavior of IL electrolytes is the PNP equation, which is a partial differential equation that describes the transport of ions in an electrolyte solution. The PNP equation takes into account the electrostatic interactions between ions, the concentration gradients of the ions, and the external electric field. The PNP equation can be solved using numerical methods, such as finite difference or finite element methods, to obtain the spatial distribution of ions in the electrolyte (233).
- *Molecular Dynamics (MD) Simulation*: Another theoretical model used to study the behavior of IL electrolytes in lithium batteries is MD simulation. MD is a computational method that simulates the movement of atoms and molecules over a period of

DOI: 10.1201/9781003531821-4

time, using classical mechanics. In MD simulation, the interactions between atoms and molecules are described by a potential energy function, which can be parameterized using experimental data or quantum mechanical calculations. MD simulation can be used to study the structure and dynamics of IL electrolytes, as well as the interactions between IL electrolytes and electrode surfaces (234–239).

As a case study, a key challenge in the utilization of polymerized ionic liquids (polyILs) as electrolytes in energy storage devices is the observed reduction in anion diffusivities compared to their parent ionic liquids (ILs). This disparity in ion transport behavior between the two systems has prompted the need for a deeper understanding of the underlying mechanisms governing ion transport in these materials. In the following case study, we explore a comparative analysis of ion transport in ILs and polyILs, leveraging the insights gained from molecular dynamics simulations. By exploring the distinct ion transport mechanisms in these materials, we aim to provide valuable guidance for the design and optimization of polyIL electrolytes with enhanced performance in advanced energy storage applications. The findings from this case study offer a comprehensive understanding of the factors influencing ion diffusion in ILs and polyILs, shedding light on the challenges and opportunities associated with the utilization of polyILs in next-generation energy storage technologies (240).

Case Study 1: Comparative Analysis of Ion Transport in ILs and PolyILs

SYNOPSIS

This case study investigates the comparative assessment of ion transport dynamics in conventional ILs and polyILs. The investigation centers on unraveling the fundamental mechanisms governing ion movement in these distinct systems, shedding light on the obstacles and potentials associated with incorporating polyILs into energy storage and electrochemical devices.

OBJECTIVE

The primary aim of this research is to compare and analyze ion transport mechanisms in ILs and polyILs through the lens of molecular dynamics simulations. By scrutinizing the factors influencing ion diffusion in these materials, the study seeks to offer insights that can shape the design and enhancement of polyIL electrolytes for superior performance in energy storage applications.

METHODOLOGY

- Utilized molecular dynamics simulations to explore ion transport in ILs and polyILs.
- Examined ion transport mechanisms in ILs, with a focus on ion association dynamics and structural relaxation.
- Explored the distinctive aspects of ion transport in polyILs, particularly emphasizing the impact of polymer chain dynamics and anion diffusion within cationic polymer cages.

FINDINGS

- Revealed that in ILs, ion transport is intricately linked to ion association dynamics and structural relaxation, affecting the rate of ion diffusion.
- Uncovered that in polyILs, anion diffusion is predominantly governed by anions hopping between cationic polymer cages, with the "trap time" dictating the pace of anion transport.
- Emphasized the influence of factors like free volume fraction, polymer chain oscillations, and chain translation speed on ion diffusivities in poly-ILs across different temperatures.

CONCLUSION

The comparative exploration of ion transport in ILs and polyILs yields valuable insights into the distinct mechanisms dictating ion diffusion in these materials. By discerning the variations in ion transport behavior, researchers can refine the design of polyIL electrolytes for enhanced performance in energy storage and electrochemical devices.

KEY INSIGHTS

- Molecular dynamics simulations serve as a potent tool for investigating ion transport mechanisms in ILs and polyILs.
- The unique characteristics of polyILs, including polymer chain dynamics and cationic polymer cages, significantly impact ion diffusion behavior.
- Lessons from this study can steer the development of advanced polyIL electrolytes with improved ion transport properties for cutting-edge energy storage applications.
- Insights from this study can guide the development of advanced polyIL electrolytes with enhanced ion transport properties for next-generation energy storage applications.

This case study underscores the importance of understanding and comparing ion transport in traditional ILs and emerging polyILs, providing valuable insights for the design and optimization of polyIL electrolytes in advanced energy storage systems.

- *Density Functional Theory (DFT)*: DFT is a quantum mechanical method that can calculate the electronic structure and energetics of molecules and materials. DFT can be used to study the adsorption of ions and molecules on electrode surfaces, as well as the formation and decomposition of solid electrolyte interphase (SEI) layers (241–245). In this section, we delve into the realm of computational analysis to unlock the potential of alicyclic ILs as electrolytes in lithium metal batteries. The study presented here aims to uncover the intricate electrochemical stability, charge transfer moments, and ion interactions within these ILs, paving the way for optimized performance in battery applications (246–248).

As we navigate through the methodology employed to dissect the properties of alicyclic ILs, we uncover key findings that shed light on their suitability as electrolytes for lithium metal batteries. By understanding the complex interplay between anions, cations, lithium ions, and ion pairs within these IL electrolytes, we gain valuable insights into enhancing battery performance and longevity.

Join us on this journey as we leverage computational modeling to unravel the behavior and potential of alicyclic ILs in lithium metal battery electrolytes.

Case Study 2. Computational Analysis of Alicyclic ILs in Lithium Metal Battery Electrolytes

OVERVIEW

This case study delves into the computational exploration of alicyclic ILs as electrolytes in lithium metal batteries. The research focuses on understanding the electrochemical stability windows, charge transfer moments, and ion interactions within these ILs to enhance their performance in battery applications.

OBJECTIVE

The primary objective of the study is to investigate the electrochemical stability, bulk properties, and ion interactions of alicyclic ILs to optimize their utilization as electrolytes in lithium metal batteries. By employing computational approaches, the aim is to enhance the efficiency and reliability of these ILs in battery systems.

METHODOLOGY

- Utilized computational modeling to analyze the electrochemical stability windows of alicyclic ILs as lithium metal battery electrolytes.
- Investigated how the ionic structure of alicyclic ILs influences their bulk properties and charge transfer moments.
- Explored the interactions between anions, cations, lithium ions, and ion pairs within alicyclic IL electrolytes for lithium metal batteries.

FINDINGS

- Identified the electrochemical stability windows of alicyclic ILs, providing insights into their suitability as electrolytes for lithium metal batteries.
- Analyzed the charge transfer moments of alicyclic ILs to understand their charge distribution and transfer behavior in battery systems.
- Investigated the complex interactions between anions, cations, lithium ions, and ion pairs within alicyclic IL electrolytes, shedding light on their impact on battery performance, see Figure 3.

FIGURE 3 AIM picture for RmAzp+NTf2+Li. The symbols for every atom except hydrogen are indicated on the corresponding atom.

CONCLUSION

The computational analysis of alicyclic ILs in lithium metal battery electrolytes offers valuable insights into their electrochemical behavior, bulk properties, and ion interactions. By understanding these aspects, researchers can tailor the design and composition of alicyclic ILs to enhance their functionality and stability in lithium battery applications.

KEY TAKEAWAYS

Computational modeling provides a powerful tool for studying the electrochemical behavior of alicyclic ILs in lithium metal batteries. Understanding the charge transfer moments and ion interactions within these ILs is crucial for optimizing their performance. Insights from this study can guide the development of advanced electrolytes for more efficient and reliable lithium metal battery systems.

This case study highlights the significance of computational analysis in elucidating the behavior and properties of alicyclic ILs as electrolytes in lithium metal batteries, paving the way for enhanced battery performance and longevity.

- *Monte Carlo (MC) Simulation*: This is a statistical method that can be used to simulate the behavior of a system with many interacting particles. MC simulation can be used to study the thermodynamics and kinetics of ion transport and reactions in IL electrolytes (249–252).
- *Continuum Models*: These models can be used to describe the transport of ions and electrons in the electrolyte and electrode materials. Continuum models are based on the laws of conservation of mass, momentum, and energy. They can be used to describe the behavior of a system at a larger scale than molecular simulations (253–257).
- *Hybrid Models*: These models combine multiple computational approaches, such as MD and continuum models, to capture both the detailed molecular interactions at the interface and the macroscopic behavior of the electrode-IL system. Hybrid models can provide a more complete understanding of the interface by incorporating both electronic and ionic interactions, as well as the effects of external factors, such as temperature and pressure (258).

Experimental techniques, such as electrochemical measurements, spectroscopy, and microscopy, have been used to study the behavior of IL electrolytes in lithium batteries.

- *Electrochemical Measurements*: Measurements such as cyclic voltammetry and electrochemical impedance spectroscopy can be used to measure the electrochemical properties of the electrolyte and electrode materials, including their capacitance, resistance, and diffusion coefficient. Cyclic voltammetry (CV) is a valuable electrochemical method utilized in battery research to assess the electrochemical characteristics of materials used in batteries. It involves analyzing the current response of a redox active solution to a linear potential sweep, providing insights into redox processes, energy levels, and electronic-transfer kinetics. By sweeping the electrode potential linearly over time and measuring the resulting current flow in an electrochemical cell, CV yields essential electrochemical data about the material being studied. This technique is widely applied in diverse fields like analytical chemistry, materials science, and electrochemistry for both research purposes and practical applications, offering crucial information on electroactive species behavior, electrochemical kinetics, diffusion coefficients, concentration analysis, and electrode surface properties.

Electrochemical impedance spectroscopy (EIS) is a technique used to measure the impedance of a system as a function of the AC potentials frequency. It is a powerful method that provides insights into the behavior of complex electrochemical systems by isolating and distinguishing the influence of various physical and chemical phenomena. EIS is widely employed in diverse fields, including batteries, catalysis, corrosion processes, semiconductor interfaces, and ion diffusion across membranes. EIS measures the resulting current response as a sine wave superimposed on the DC current, providing valuable information about the system's impedance as a function of frequency. EIS is used in electrochemical measurements to characterize the behavior of electrochemical systems, study electrode kinetics, analyze mass transport phenomena, and evaluate the performance of protective coatings against corrosion1. It is a versatile technique that allows researchers to rapidly characterize electrochemical systems, study processes from high to low frequencies, and optimize system behavior at different operating points, such as different states of charge in batteries. Overall, EIS is a valuable tool in electrochemical measurements, providing detailed insights into the electrochemical behavior of systems, enabling the characterization of various electrochemical processes, and offering opportunities for system optimization and performance enhancement in a wide range of applications. Some other electrochemical techniques used to measure the electrochemical properties of electrolyte and electrode materials include: chronoamperometry, chronopotentiometry, potentiostatic intermittent titration technique (PITT), and linear sweep voltammetry.

- *Spectroscopy Techniques*: Techniques such as infrared and Raman spectroscopy can be used to study the chemical structure and composition of the electrolyte and electrode materials. Spectroelectrochemistry (SEC) combines electrochemistry and spectroscopy techniques to offer enhanced insights compared to using them separately. It enables real-time characterization of electrogenerated species. Among the four spectroscopy techniques with high potential in SEC, IR-SEC, and Raman-SEC are highlighted for their ability to analyze the chemical structure and composition of electrolyte and electrode materials. NMR-SEC is also recognized as a valuable method for understanding electrochemical systems.

- *Microscopy Techniques*: Microscopy techniques, including scanning electron microscopy (SEM) and transmission electron microscopy (TEM), are valuable tools used to visualize the morphology and structure of electrode materials and solid electrolyte interphase (SEI) layers. These techniques provide high-resolution imaging capabilities that allow researchers to examine the surface morphology, structure, and composition of electrode materials at the micro- and nano-scale levels. SEM is particularly useful for studying the surface topography and elemental composition of materials, while TEM provides detailed insights into the internal structure and composition of materials at the atomic level. By utilizing SEM and TEM, researchers can gain a comprehensive understanding of the physical characteristics and properties of electrode materials and SEI layers, aiding in the development and optimization of advanced energy storage devices (259).

In summary, studying the behavior of IL electrolytes in lithium batteries is a complex and interdisciplinary field. By combining theoretical models, computational methods, and experimental techniques, researchers can gain a comprehensive understanding of the behavior of IL electrolytes in lithium batteries. This can lead to the design of new IL electrolytes with improved performance, such as higher conductivity, stability, and safety, as well as the optimization of electrode materials and system designs to maximize battery performance and lifetime.

Validation, Management, and Reproducibility

5

HOW TO VALIDATE THE ACCURACY OF COMPUTATIONAL MODELS FOR THE ELECTRODE IONIC LIQUID INTERFACE?

Validating the accuracy of computational models for the electrode-ionic liquid interface is an important step in ensuring that the models represent the real-world system. Here are a few ways that researchers can validate the accuracy of computational models:

- *Comparison with Experimental Measurements*: Researchers validate computational models by comparing their predictions with experimental measurements of the electrode-ionic liquid interface using techniques like X-ray photoelectron spectroscopy (XPS), infrared spectroscopy (IR), or scanning tunneling microscopy (STM). This comparison helps assess the accuracy of the computational model and pinpoint areas for enhancement based on the agreement with experimental data (260–263).
- *Sensitivity analysis*: This is a method utilized to evaluate how sensitive a model is to changes in input parameters and assumptions. By adjusting these parameters and observing the resulting impact on the model's predictions, researchers can determine which parameters significantly influence the model's accuracy and may require further refinement or detailed characterization (264,265).
- *Consistency with Known Physical Principles*: Another way to validate a computational model is to ensure that it is consistent

DOI: 10.1201/9781003531821-5

with known physical principles (266,267). For example, the model should conserve energy and momentum, and it should satisfy fundamental laws such as the laws of thermodynamics (268). By verifying that the computational model upholds these physical constraints and principles, researchers can increase confidence in the model's accuracy and reliability. This type of validation complements the comparison to experimental data, as it ensures the model is not only predictive, but also grounded in the underlying physical reality. Validating a computational model against physical principles is an important step in the overall validation process, as it helps identify potential flaws or inconsistencies in the model formulation, prior to comparing it to empirical observations.

- *Comparison with Other Models*: Researchers may also validate their computational models by comparing them with other models validated in the literature (269). By comparing the predictions of different models, researchers can identify areas of agreement and disagreement and gain insights into the limitations and uncertainties of the models. This may be achieved by comparing their predictions, analyzing the consistency in outcomes, and assessing the alignment or discrepancies in results.

  Researchers compare the outputs of different computational models to determine where they agree or disagree in their predictions. They evaluate the consistency in predictions across various models, with consistent results indicating agreement and inconsistencies highlighting disagreements. Researchers also analyze the alignment or discrepancies in results concerning specific parameters or assumptions to identify factors contributing to agreement or disagreement between the models.

- *Reproducibility*: Another critical aspect of validating computational models is reproducibility. The model should be well-documented and the code should be made available to other researchers, so that others can reproduce the results and validate the model independently (270–272).

- *Benchmarking*: Finally, researchers may validate their computational models by benchmarking them against known or reference systems (273). For example, researchers may use standard test cases or reference systems to validate the accuracy of their models. By benchmarking the model against a known system, researchers can assess the accuracy and reliability of the model and identify areas for improvement.

A common benchmarking system for validating computational models of the electrode-ionic liquid interface is the graphite-electrolyte interface. The graphite electrode is a widely used electrode material in lithium-ion batteries, and its interface with the electrolyte has been extensively studied experimentally and computationally (37,228,239,274,275). These studies have demonstrated the importance of accurately modeling the interactions between the graphite electrode and the electrolyte, including the effects of surface roughness, solvent molecules, and charge transfer processes. They have also highlighted the limitations and uncertainties of the models and identified areas for improvement.

Experimental measurements of the graphite-electrolyte interface can be obtained using techniques such as X-ray photoelectron spectroscopy (XPS), infrared spectroscopy (IR), and scanning tunneling microscopy (STM). These measurements can be used to validate the accuracy of computational models, by comparing the model predictions with the experimental data.

Computational models of the graphite-electrolyte interface typically involve molecular dynamics simulations, in which the atomic interactions and dynamics are simulated over a while. The accuracy of the model can be assessed by comparing the simulation results with experimental measurements, as well as with other computational models that have been validated in the literature. Sensitivity analysis can also be used to identify which parameters are most critical to the accuracy of the model, and which may need to be refined or better characterized.

The graphite-electrolyte interface provides a well-defined system for benchmarking computational models and has been used in numerous studies for this purpose (276–278).

In conclusion, validating the accuracy of computational models for the electrode-ionic liquid interface involves a comprehensive approach. This includes integrating experimental measurements, sensitivity analysis, adherence to physical principles, comparison with existing models, ensuring reproducibility, and benchmarking. By employing these diverse strategies, researchers can enhance their confidence in the precision and dependability of their models. This, in turn, enables the development of more efficient strategies for enhancing the performance and stability of lithium-ion batteries.

SOME COMMON CHALLENGES RESEARCHERS FACE WHEN VALIDATING COMPUTATIONAL MODELS

Validating computational models can be challenging, and researchers may face several common challenges. Here are a few examples:

- *Limited Experimental Data*: The main challenge in validating computational models is the scarcity of experimental data for comparison, particularly in the context of the electrode-ionic liquid interface. Obtaining experimental measurements for this interface is often challenging, expensive, and prone to errors and uncertainties. Consequently, researchers may have to work with a restricted amount of experimental data for validation, which can hinder the thorough assessment of the model's accuracy and reliability.
- *Variability of Experimental Data*: Even when experimental data is available, it can be subject to variability and uncertainty. For example, measurements of the electrode-ionic liquid interface can be affected by factors such as surface roughness, impurities, and sample preparation, which can lead to variability in the data. This can make it challenging to validate the accuracy of the model and to identify the sources of error and uncertainty.
- *Complexity of the System*: The electrode-ionic liquid interface is a complex system with many interacting components, including the electrode material, the ionic liquid electrolyte, and the interface between them. Modeling this system accurately requires a high level of computational complexity, and may require simplifying assumptions and approximations that can limit the accuracy of the model.
- *Limitations of the Model*: Computational models for the electrode-ionic liquid interface often rely on simplifications and approximations, potentially overlooking crucial phenomena and interactions at the interface. These models may omit factors like solvent molecules, small ions, or charge transfer mechanisms, impacting their accuracy and relevance. Researchers must carefully consider the model's constraints and the assumptions underlying it to interpret results accurately.
- *Computational Resources*: Validating computational models can be computationally intensive and may require significant computational

resources, especially for large-scale systems or for models that incorporate multiple levels of detail and complexity. This can make it challenging for researchers to run and validate the model within a reasonable time frame or with available computational resources.

- *Model Transferability*: Another challenge in validating computational models is their transferability to other systems or conditions. Models that are validated for a specific system or set of conditions may not be applicable for other systems or under different conditions, which can limit their utility. Researchers need to carefully assess the transferability of the model and its assumptions when applying it to new systems or conditions.

Validating computational models for the electrode-ionic liquid interface requires careful consideration of the limitations and challenges involved, as well as the strengths and weaknesses of the model and the available experimental data. By addressing these challenges and refining the models over time, researchers can develop more accurate and reliable models that can guide the development of new materials and devices for lithium-ion batteries.

TECHNIQUES FOR MANAGING AND ANALYZING LARGE AMOUNTS OF DATA GENERATED BY SIMULATIONS

Managing and analyzing large amounts of data generated by simulations is a common challenge in computational modeling. Here are a few techniques that can help researchers manage and analyze the data efficiently:

- *Parallel Processing*: This is a technique used to distribute the computational workload across multiple processors or nodes (279,280). By running simulations in parallel, researchers can reduce the time required to generate the data and increase the efficiency of the simulations. Parallel processing can be implemented using specialized software and hardware, such as graphical processing units (GPUs) and high-performance computing (HPC) clusters.

  GPUs are designed with parallel processing architecture, allowing them to handle resource-intensive tasks efficiently. They consist of multiple cores that can perform extensive calculations simultaneously, making them well-suited for tasks like ML model training, data mining operations, and high-resolution graphics rendering. GPUs can integrate multiple units to enhance processing potential, consume less memory, and execute tasks faster due to their parallel processing nature.

HPC clusters leverage both CPUs and GPUs to perform diverse operations simultaneously. While CPUs handle serial processing for various applications and the operating system within the cluster, GPUs excel at parallel processing for massive external workloads like ML model training and data mining. HPC clusters can be designed without GPUs, but including GPUs significantly boosts the system's performance. CPUs are essential for running an HPC system, and GPUs further enhance the processing power, especially for resource-intensive tasks.

- *Data Compression*: This is a technique used to reduce the size of the data generated by the simulations without losing important information (281,282).

  Data compression is the process of encoding information using fewer bits than the original representation. This technique, also known as source coding or bit-rate reduction, aims to reduce the size of data without losing its essential content.

  There are two main types of data compression:

  Lossless Compression: This method reduces bits by eliminating statistical redundancy without losing any information. The original data can be perfectly reconstructed from the compressed form.

  Lossy Compression: This approach removes unnecessary or less important information, resulting in a smaller file size but with some loss of data accuracy or detail.

  For example, researchers may use lossless compression techniques, such as gzip or bzip2, to compress data files without losing any information (283–286). Alternatively, researchers may use lossy compression techniques, such as JPEG or MP3, to reduce the size of data files while sacrificing some level of accuracy or detail (287,288).

  The device that performs the data compression is referred to as an encoder, while the one that reverses the process (decompression) is known as a decoder.

- *Data Visualization*: This is a technique used to represent the data generated by the simulations in a visual format that is easy to interpret and analyze (289,290). For example, researchers may use plots, graphs, or 3D visualizations to represent the data and identify patterns and trends Figure 4. Data visualization can be implemented using specialized software tools, such as MATLAB, Python, or ParaView.

The choice of data visualization technique depends on factors such as the type of data, the research question, and the target audience. Principles of effective data visualization, such as clarity, conciseness, and appropriate use of color and scale, should be considered when creating visualizations.

- *Machine Learning*: This is a technique used to analyze large amounts of data and identify patterns and trends automatically (291,292). For example, researchers may use clustering or regression algorithms to analyze the data generated by the simulations and identify correlations between variables or features (224,232,293–295). Machine learning can be implemented using specialized software tools, such as scikit-learn (296) or TensorFlow.

The choice of machine learning technique depends on the type of data, the research question, and the desired insights.

Effective use of machine learning requires careful data preprocessing, model selection, and validation to ensure the reliability and accuracy of the results.

Some advantages of using machine learning over traditional methods of data analysis include:

In summary, machine learning excels at uncovering intricate patterns and relationships in data that traditional statistical methods may overlook. Its scalability allows for efficient processing of large datasets, making it ideal for big data analysis. Machine learning models can adjust to data changes, ensuring ongoing accuracy. By minimizing human biases, machine learning delivers more objective predictions. It automates decision-making, enhancing speed and reducing manual intervention. With the ability to analyze diverse data sources, machine learning enhances forecasting and decision-making accuracy. Its flexibility and adaptability across various domains make it a versatile tool for a wide range of applications.

The limitation of machine learning is its dependence on the quality and quantity of data. Insufficient or biased data can lead to inaccurate results, impacting the effectiveness and reliability of machine learning algorithms.

- *Data Management Software*: This is a tool used to organize, store, and retrieve large amounts of data generated by simulations (297). For example, researchers may use specialized software tools, such as MongoDB (298) or Cassandra (299), to manage and query large datasets efficiently and effectively. Data management software can

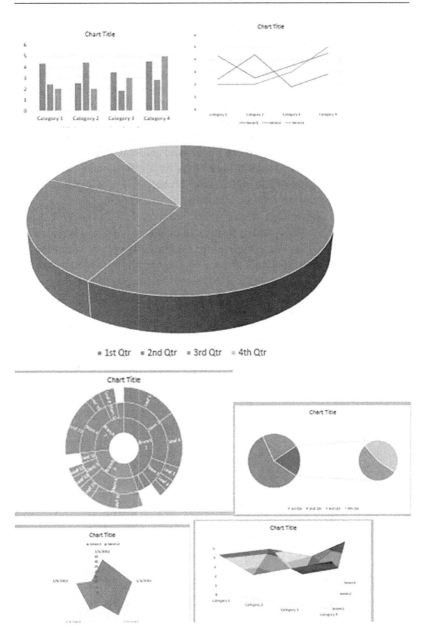

FIGURE 4   Some types of charts for data visualization.

also be used to automate routine tasks, such as data backup and archiving (300).

In contemporary enterprises, data management software is essential. For the purpose of managing, organizing, and storing enormous volumes of data, data management services offer an organized framework. Databases are the foundation of everything from keeping track of inventory to maintaining customer data. Data management software effectively stores large amounts of data, arranging it into rows, tables, and columns. This structured organization facilitates efficient data retrieval, allowing users to quickly obtain specific information through the use of queries. The software also enables data manipulation, empowering users to add, edit, or remove data, ensuring the accuracy and currency of the database. To safeguard sensitive information, data management services employ security features such as encryption, authorization, and authentication, preventing breaches or unwanted access. Additionally, these tools offer robust data recovery and backup capabilities, routinely backing up data to ensure the continuation and integrity of the information, even in the event of system failures.

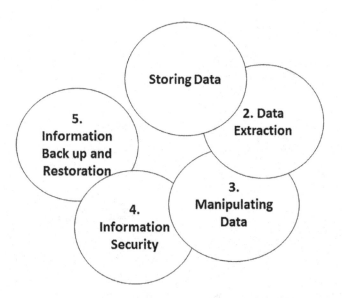

FIGURE 5 Five essential functions in data management.

By addressing these core database management functions, data management software plays a crucial role in supporting businesses, enabling them to effectively leverage their data to drive informed decision-making and foster innovation (301).

- *Collaborative Platforms*: Collaborative platforms are online tools that allow researchers to share and analyze data collaboratively (302). For example, platforms such as GitHub or GitLab can be used to share code and data between researchers (300,303). While platforms such as Jupyter Notebooks or Google Colab can be used to run and analyze simulations in a collaborative environment (304,305).

HOW TO IMPROVE THE REPRODUCIBILITY OF COMPUTATIONAL MODELS

Improving the reproducibility of computational models is an important goal for researchers, as it enables other researchers to validate and build upon their work, and ensures that the results are reliable and accurate. Here are a few strategies that can help improve the reproducibility of computational models:

- *Documenting the Methodology*: Researchers should document the methodology used to develop and validate the computational model, including the software tools, input parameters, and simulation protocols. This documentation should be detailed and transparent, and should be made available to other researchers through published papers, technical reports, or online repositories.
- *Version Control*: Version control is a technique used to track changes to the code and data used in the computational model, and to ensure that the results can be reproduced precisely. Researchers should use version control tools, such as Git or SVN, to track changes to the code and data, and to document the history of the model development (306).
- *Sharing the Code*: Sharing the code used to develop the computational model is an important step in improving reproducibility, as it allows other researchers to reproduce the results and build upon the work. Researchers should make their code available in online

repositories, such as GitHub or GitLab, and should ensure that the code is well-documented and easy to use.

- *Providing Input Data and Results*: This strategy used in the computational model is another important step in improving reproducibility. Researchers should make their input data and results available in online repositories, such as Figshare or Zenodo, and should ensure that the data is well-organized and annotated (307). This will enable other researchers to reproduce the results and build upon the work.

- *Using Open-Source Software*: Using open-source software tools can also help improve the reproducibility of computational models, as it allows other researchers to use and modify the software without restrictions. Researchers should use open-source software tools and should ensure that they use well-established and widely-used libraries and frameworks.

- *Conducting Sensitivity Analysis*: This is an important step in improving the reproducibility of computational models, as it enables researchers to identify the input parameters that are most critical to the accuracy of the model (308,309). Sensitivity analysis should be conducted using a range of input parameter values, and the results should be documented and reported in publications.

- *Peer Review*: Finally, This is an important step in improving the reproducibility of computational models, as it allows other researchers to evaluate the validity and accuracy of the model. Researchers should submit their work to peer-reviewed journals or conferences, and should ensure that their work is subject to rigorous peer review.

Overall, improving the reproducibility of computational models requires a combination of strategies, including documenting the methodology, using version control, sharing the code and data, using open-source software, conducting sensitivity analysis, and peer review. By following these strategies, researchers can ensure that their work is reliable, accurate, and useful to the broader scientific community.

Conclusion

6

Ionic liquids have unique properties that make them suitable for use as electrolytes in lithium batteries. However, integrating them into existing battery designs may be challenging and require modifications. Continued research is needed to optimize the properties of ionic liquids and develop new types with improved characteristics. The interaction between cations and anions in ionic liquids is a delicate balance of forces. Researchers are exploring new battery designs or modifications to accommodate ionic liquid electrolytes.

The ion transport mechanism in ionic liquid electrolytes differs from traditional organic solvent-based electrolytes. High ionic conductivity and fast ion transport are crucial for high power output in applications like electric vehicles. The viscosity of the ionic liquid electrolyte affects ion transport and diffusion in batteries. Researchers are exploring ways to modify the ionic liquid structure or add co-solvents to reduce viscosity. Validating computational models for the electrode-ionic liquid interface requires experimental measurements, sensitivity analysis, consistency with physical principles, comparison with other models, reproducibility, and benchmarking. Careful consideration of limitations, challenges, strengths, weaknesses, and available experimental data is necessary for validating these models.

Managing and analyzing large amounts of data generated by simulations is a common challenge in computational modeling. To improve the reproducibility of computational models, researchers should document the methodology, use version control, share the code, conduct sensitivity analysis, and undergo peer review.

Improving the reproducibility of computational models is crucial for researchers. Strategies to achieve this include documenting the methodology, using version control, sharing the code and data, using open-source software, conducting sensitivity analysis, and undergoing peer review. These measures ensure that the work is reliable, accurate, and beneficial to the scientific community.

5 Conclusion

Bibliography

1. Chen T, Jin Y, Lv H, Yang A, Liu M, Chen B, et al. Applications of lithium-ion batteries in grid-scale energy storage systems. *Trans Tianjin Univ* [Internet]. 2020;26(3):208–17. Available from: https://doi.org/10.1007/s12209-020-00236-w
2. Chayambuka K, Mulder G, Danilov DL, Notten PHL. From Li-ion batteries toward Na-ion chemistries: Challenges and opportunities. *Adv Energy Mater.* 2020;10(38):1–11.
3. Dunn B, Kamath H, Tarascon JM. Electrical energy storage for the grid: A battery of choices. *Science (80-).* 2011;334(6058):928–35.
4. Lakshmi GS, Olena R, Divya G, Oleksandr R. Battery energy storage technologies for sustainable electric vehicles and grid applications. *J Phys Conf Ser.* 2020;1495(1):12014.
5. Maciej S, Becker FG, Cleary M, Team RM, Holtermann H, The D, et al. Synteza i aktywność biologiczna nowych analogów tiosemikarbazonowych chelatorów żelaza. G. Balint, Antala B, Carty C, Mabieme J-MA, Amar IB, Kaplanova A, editors. *Uniw śląski* [Internet]. 2013 [cited 2023 Jun 7];7(1):343–54.
6. Keçili R, Arli G, Hussain CM. Future of analytical chemistry with graphene. *Compr Anal Chem.* 2020;91:355–89.
7. Zhang ZJ, Ramadass P, Fang W. Safety of lithium-ion batteries. In: *Lithium-Ion Batteries—Advances and Applications.* Pistoia G, editor. Philadelphia, PA: Elsevier; 2014 Jan 1. p. 409–35.
8. Hamed MM, El-Tayeb A, Moukhtar I, El Dein AZ, Abdelhameed EH. A review on recent key technologies of lithium-ion battery thermal management: External cooling Systems. *Results Eng* [Internet]. 2022;16(July):100703. Available from: https://doi.org/10.1016/j.rineng.2022.100703
9. Yoshizawa H. Secondary batteries – Lithium rechargeable systems – Lithium-ion I Lithium vanadium oxide/niobium oxide batteries. *Encycl Electrochem Power Sources.* 2009 Jan 1;368–74.
10. Xiong H, Dufek EJ, Gering KL. 2.20 Batteries. *Compr Energy Syst.* 2018 Jan 1;2–5:629–62.
11. Zeng X, Li M, Abd El-Hady D, Alshitari W, Al-Bogami AS, Lu J, et al. Commercialization of lithium battery technologies for electric vehicles. *Adv Energy Mater.* 2019;9(27):1–25.
12. Tabor DP, Roch LM, Saikin SK, Kreisbeck C, Sheberla D, Montoya JH, et al. Accelerating the discovery of materials for clean energy in the era of smart automation. *Nat Rev Mater.* 2018;3(5):5–20.
13. Li M, Lu J, Chen Z, Amine K. 30 years of lithium-ion batteries. *Adv Mater.* 2018;30(33):1–24.

14. Ding Y, Cano ZP, Yu A, Lu J, Chen Z. Automotive Li-ion batteries: Current status and future perspectives. *Electrochem Energy Rev* [Internet]. 2019;2(1). Available from: https://doi.org/10.1007/s41918-018-0022-z
15. Jiang X, Chen Y, Meng X, Cao W, Liu C, Huang Q, et al. The impact of electrode with carbon materials on safety performance of lithium-ion batteries: A review. *Carbon N Y* [Internet]. 2022 May;191:448–70. Available from: https://linkinghub.elsevier.com/retrieve/pii/S0008622322000884
16. Shan T, Wang Z, Zhu X, Wang H, Zhou Y, Wang Y, et al. Explosion behavior investigation and safety assessment of large-format lithium-ion pouch cells. *J Energy Chem* [Internet]. 2022 Sep;72:241–57. Available from: https://linkinghub.elsevier.com/retrieve/pii/S209549562200208X
17. Wang Y-W, Shu C-M. Energy generation mechanisms for a Li-ion cell in case of thermal explosion: A review. *J Energy Storage* [Internet]. 2022 Nov;55:105501. Available from: https://linkinghub.elsevier.com/retrieve/pii/S2352152X22014931
18. Qi C, Zhu Y-L, Gao F, Wang S-C, Yang K, Jiao Q-J. Safety Analysis of lithium-ion battery by rheology-mutation theory coupling with fault tree method. *J Loss Prev Process Ind* [Internet]. 2017 Sep;49:603–11. Available from: https://linkinghub.elsevier.com/retrieve/pii/S0950423017305296
19. Hu G, Huang P, Bai Z, Wang Q, Qi K. Comprehensively analysis the failure evolution and safety evaluation of automotive lithium ion battery. *eTransportation* [Internet]. 2021 Nov;10:100140. Available from: https://linkinghub.elsevier.com/retrieve/pii/S2590116821000382
20. Deng J, Bae C, Denlinger A, Miller T. Electric vehicles batteries: Requirements and challenges. *Joule* [Internet]. 2020 Mar;4(3):511–5. Available from: https://linkinghub.elsevier.com/retrieve/pii/S254243512030043X
21. Rangarajan S, Sunddararaj SP, Sudhakar A, Shiva CK, Subramaniam U, Collins ER, et al. Lithium-ion batteries—The crux of electric vehicles with opportunities and challenges. *Clean Technol* [Internet]. 2022 Sep 21;4(4):908–30. Available from: https://www.mdpi.com/2571-8797/4/4/56
22. Lawder MT, Suthar B, Northrop PWC, De S, Hoff CM, Leitermann O, et al. Battery energy storage system (BESS) and battery management system (BMS) for grid-scale applications. *Proc IEEE* [Internet]. 2014 Jun;102(6):1014–30. Available from: http://ieeexplore.ieee.org/document/6811152/
23. Mayyas A, Chadly A, Amer ST, Azar E. Economics of the Li-ion batteries and reversible fuel cells as energy storage systems when coupled with dynamic electricity pricing schemes. *Energy* [Internet]. 2022 Jan;239:121941. Available from: https://linkinghub.elsevier.com/retrieve/pii/S0360544221021897
24. Bae S-Y, Shim E-G, Kim D-W. Effect of ionic liquid as a flame-retarding additive on the cycling performance and thermal stability of lithium-ion batteries. *J Power Sources* [Internet]. 2013 Dec;244:266–71. Available from: https://linkinghub.elsevier.com/retrieve/pii/S0378775313001572
25. Niu H, Wang L, Guan P, Zhang N, Yan C, Ding M, et al. Recent advances in application of ionic liquids in electrolyte of lithium ion batteries. J Energy Storage [Internet]. 2021 Aug;40:102659. Available from: https://linkinghub.elsevier.com/retrieve/pii/S2352152X21003984
26. Xiang J, Wu F, Chen R, Li L, Yu H. High voltage and safe electrolytes based on ionic liquid and sulfone for lithium-ion batteries. *J Power Sources* [Internet]. 2013 Jul;233:115–20. Available from: https://linkinghub.elsevier.com/retrieve/pii/S0378775313001808

27. Balducci A. Ionic liquids in lithium-ion batteries. *Top Curr Chem* [Internet]. 2017 Apr 2;375(2):20. Available from: http://link.springer.com/10.1007/s41061-017-0109-8

28. Tang X, Lv S, Jiang K, Zhou G, Liu X. Recent development of ionic liquid-based electrolytes in lithium-ion batteries. *J Power* Sources [Internet]. 2022 Sep;542:231792. Available from: https://linkinghub.elsevier.com/retrieve/pii/S0378775322007820

29. Kühnel R-S, Böckenfeld N, Passerini S, Winter M, Balducci A. Mixtures of ionic liquid and organic carbonate as electrolyte with improved safety and performance for rechargeable lithium batteries. *Electrochim Acta* [Internet]. 2011 Apr;56(11):4092–9. Available from: https://linkinghub.elsevier.com/retrieve/pii/S0013468611001903

30. Tan S, Ji YJ, Zhang ZR, Yang Y. Recent progress in research on high-voltage electrolytes for lithium-ion batteries. *ChemPhysChem* [Internet]. 2014 Jul 21;15(10):1956–69. Available from: https://onlinelibrary.wiley.com/doi/10.1002/cphc.201402175

31. Hyun WJ, Thomas CM, Luu NS, Hersam MC. Layered heterostructure ionogel electrolytes for high-performance solid-state lithium-ion batteries. *Adv Mater* [Internet]. 2021 Apr 17;33(13):2007864. Available from: https://onlinelibrary.wiley.com/doi/10.1002/adma.202007864

32. Liu K, Wang Z, Shi L, Jungsuttiwong S, Yuan S. Ionic liquids for high performance lithium metal batteries. *J Energy Chem* [Internet]. 2021 Aug;59:320–33. Available from: https://linkinghub.elsevier.com/retrieve/pii/S2095495620307671

33. Cavers H, Molaiyan P, Abdollahifar M, Lassi U, Kwade A. Perspectives on improving the safety and sustainability of high voltage lithium-ion batteries through the electrolyte and separator region. *Adv Energy Mater* [Internet]. 2022 Jun 6;12(23):2200147. Available from: https://onlinelibrary.wiley.com/doi/10.1002/aenm.202200147

34. Zhou W, Zhang M, Kong X, Huang W, Zhang Q. Recent advance in ionic-liquid-based electrolytes for rechargeable metal-ion batteries. *Adv Sci* [Internet]. 2021 Jul 2;8(13):2004490. Available from: https://onlinelibrary.wiley.com/doi/10.1002/advs.202004490

35. Francis CFJ, Kyratzis IL, Best AS. Lithium-ion battery separators for ionic-liquid electrolytes: A review. *Adv Mater* [Internet]. 2020 May 20;32(18):1904205. Available from: https://onlinelibrary.wiley.com/doi/10.1002/adma.201904205

36. Manthiram A, Yu X, Wang S. Lithium battery chemistries enabled by solid-state electrolytes. *Nat Rev Mater* [Internet]. 2017 Feb 14;2(4):16103. Available from: https://www.nature.com/articles/natrevmats2016103

37. Yamada Y, Wang J, Ko S, Watanabe E, Yamada A. Advances and issues in developing salt-concentrated battery electrolytes. *Nat Energy* [Internet]. 2019 Mar 11;4(4):269–80. Available from: https://www.nature.com/articles/s41560-019-0336-z

38. Palacín MR. Recent advances in rechargeable battery materials: A chemist's perspective. *Chem Soc Rev* [Internet]. 2009;38(9):2565. Available from: http://xlink.rsc.org/?DOI=b820555h

39. Wang X, Kerr R, Chen F, Goujon N, Pringle JM, Mecerreyes D, et al. Toward high-energy-density lithium metal batteries: Opportunities and challenges for solid organic electrolytes. *Adv Mater* [Internet]. 2020 May 21;32(18):1905219. Available from: https://onlinelibrary.wiley.com/doi/10.1002/adma.201905219

40. Deng D. Li-ion batteries: Basics, progress, and challenges. *Energy Sci Eng* [Internet]. 2015 Sep 23;3(5):385–418. Available from: https://onlinelibrary.wiley.com/doi/10.1002/ese3.95

41. Wu Y, Huang X, Huang L, Chen J. Strategies for rational design of high-power lithium-ion batteries. *Energy Environ Mater* [Internet]. 2021 Jan 25;4(1):19–45. Available from: https://onlinelibrary.wiley.com/doi/10.1002/eem2.12088

42. Li Q, Chen J, Fan L, Kong X, Lu Y. Progress in electrolytes for rechargeable Li-based batteries and beyond. *Green Energy Environ* [Internet]. 2016;1(1):18–42. Available from: https://www.sciencedirect.com/science/article/pii/S2468025716300218

43. Xu C, Yang G, Wu D, Yao M, Xing C, Zhang J, et al. Roadmap on ionic liquid electrolytes for energy storage devices. *Chem Asian J* [Internet]. 2021;16(6):549–62. Available from: https://onlinelibrary.wiley.com/doi/abs/10.1002/asia.202001414

44. Zou Q, Lu Y-C. Liquid electrolyte design for metal-sulfur batteries: Mechanistic understanding and perspective. *EcoMat* [Internet]. 2021;3(4):e12115. Available from: https://onlinelibrary.wiley.com/doi/abs/10.1002/eom2.12115

45. Ray A, Saruhan B. Application of ionic liquids for batteries and supercapacitors. *Materials (Basel)* [Internet]. 2021 May 29 [cited 2023 Jul 16];14(11):2942. Available from: https://www.mdpi.com/1996-1944/14/11/2942

46. Vogl T, Menne S, Kühnel R-S, Balducci A. The beneficial effect of protic ionic liquids on the lithium environment in electrolytes for battery applications. *J Mater Chem A* [Internet]. 2014;2(22):8258–65. Available from: https://doi.org/10.1039/C3TA15224C

47. Basile A, Hilder M, Makhlooghiazad F, Pozo-Gonzalo C, MacFarlane DR, Howlett PC, et al. Ionic liquids and organic ionic plastic crystals: Advanced electrolytes for safer high performance sodium energy storage technologies. *Adv Energy Mater* [Internet]. 2018;8(17):1703491. Available from: https://onlinelibrary.wiley.com/doi/abs/10.1002/aenm.201703491

48. Zhao J, Wilkins MR, Wang D. A review on strategies to reduce ionic liquid pretreatment costs for biofuel production. *Bioresour Technol* [Internet]. 2022;364:128045. Available from: https://www.sciencedirect.com/science/article/pii/S0960852422013785

49. No Title [Internet]. Ionic liquid market size share growth opportunities and forecast. 2020. Available from: https://www.mordorintelligence.com/industry-reports/ionic-liquid-market

50. Rogstad DT, Einarsrud M-A, Svensson AM. High-temperature performance of selected ionic liquids as electrolytes for silicon anodes in Li-ion batteries. *J Electrochem Soc* [Internet]. 2022;169(11):110531. Available from: https://doi.org/10.1149/1945-7111/ac9f78

51. Rosol ZP, German NJ, Gross SM. Solubility{,} ionic conductivity and viscosity of lithium salts in room temperature ionic liquids. *Green Chem* [Internet]. 2009;11(9):1453–7. Available from: https://doi.org/10.1039/B818176D

52. Asenbauer J, Ben Hassen N, McCloskey BD, Prausnitz JM. Solubilities and ionic conductivities of ionic liquids containing lithium salts. *Electrochim Acta* [Internet]. 2017;247:1038–43. Available from: https://www.sciencedirect.com/science/article/pii/S0013468617314706

53. Park MJ, Choi I, Hong J, Kim O. Polymer electrolytes integrated with ionic liquids for future electrochemical devices. *J Appl Polym Sci* [Internet]. 2013;129(5):2363–76. Available from: https://onlinelibrary.wiley.com/doi/abs/10.1002/app.39064

54. Jastorff B, Störmann R, Ranke J, Mölter K, Stock F, Oberheitmann B, et al. How hazardous are ionic liquids? Structure–activity relationships and biological testing as important elements for sustainability evaluation. *Green Chem* [Internet]. 2003;5(2):136–42. Available from: https://doi.org/10.1039/B211971D

55. Qi D, Liu Y, Liu Z, Zhang L, Chen X. Design of architectures and materials in in-plane micro-supercapacitors: Current status and future challenges. *Adv Mater* [Internet]. 2017;29(5):1602802. Available from: https://onlinelibrary.wiley.com/doi/abs/10.1002/adma.201602802

56. Gebresilassie Eshetu G, Armand M, Scrosati B, Passerini S. Energy storage materials synthesized from ionic liquids. *Angew Chemie Int Ed* [Internet]. 2014;53(49):13342–59. Available from: https://onlinelibrary.wiley.com/doi/abs/10.1002/anie.201405910

57. Yang G, Song Y, Wang Q, Zhang L, Deng L. Review of ionic liquids containing, polymer/inorganic hybrid electrolytes for lithium metal batteries. *Mater Des* [Internet]. 2020;190:108563. Available from: https://www.sciencedirect.com/science/article/pii/S0264127520300964

58. Yang Q, Zhang Z, Sun X-G, Hu Y-S, Xing H, Dai S. Ionic liquids and derived materials for lithium and sodium batteries. *Chem Soc Rev* [Internet]. 2018;47(6):2020–64. Available from: https://doi.org/10.1039/C7CS00464H

59. Zhang H, Zhao H, Khan MA, Zou W, Xu J, Zhang L, et al. Recent progress in advanced electrode materials{,} separators and electrolytes for lithium batteries. *J Mater Chem A* [Internet]. 2018;6(42):20564–620. Available from: https://doi.org/10.1039/C8TA05336G

60. Tansel B. Significance of thermodynamic and physical characteristics on permeation of ions during membrane separation: Hydrated radius, hydration free energy and viscous effects. *Sep Purif Technol* [Internet]. 2012;86:119–26. Available from: https://www.sciencedirect.com/science/article/pii/S138358661100637X

61. Izgorodina EI, Golze D, Maganti R, Armel V, Taige M, Schubert TJS, et al. Importance of dispersion forces for prediction of thermodynamic and transport properties of some common ionic liquids. *Phys Chem Chem Phys* [Internet]. 2014;16(16):7209–21. Available from: https://doi.org/10.1039/C3CP53035C

62. Dean PM, Pringle JM, MacFarlane DR. Structural analysis of low melting organic salts: perspectives on ionic liquids. *Phys Chem Chem Phys* [Internet]. 2010;12(32):9144–53. Available from: https://doi.org/10.1039/C003519J

63. Sanchora P, Pandey DK, Kagdada HL, Materny A, Singh DK. Impact of alkyl chain length and water on the structure and properties of 1-alkyl-3-methylimidazolium chloride ionic liquids. *Phys Chem Chem Phys* [Internet]. 2020;22(31):17687–704. Available from: https://doi.org/10.1039/D0CP01686A

64. Siqueira LJA, Ribeiro MCC. Molecular dynamics simulation of the ionic liquid N-ethyl-N,N-dimethyl-N-(2-methoxyethyl)ammoniumBis(trifluoromethanesulfonyl)imide. *J Phys Chem B* [Internet]. 2007 Oct 1;111(40):11776–85. Available from: https://doi.org/10.1021/jp074840c

65. Smith GD, Borodin O, Li L, Kim H, Liu Q, Bara JE, et al. A comparison of ether- and alkyl-derivatized imidazolium-based room-temperature ionic liquids: A molecular dynamics simulation study. *Phys Chem Chem Phys* [Internet]. 2008;10(41):6301–12. Available from: https://doi.org/10.1039/B808303G

66. Luo S, Zhang S, Wang Y, Xia A, Zhang G, Du X, et al. Complexes of ionic liquids with poly(ethylene glycol)s. *J Org Chem* [Internet]. 2010 Mar 19;75(6):1888–91. Available from: https://doi.org/10.1021/jo902521w

67. Ganapatibhotla LVNR, Zheng J, Roy D, Krishnan S. PEGylated imidazolium ionic liquid electrolytes: Thermophysical and electrochemical properties. *Chem Mater* [Internet]. 2010 Dec 14;22(23):6347–60. Available from: https://doi.org/10.1021/cm102263s

68. Tang S, Baker GA, Zhao H. Ether- and alcohol-functionalized task-specific ionic liquids: Attractive properties and applications. *Chem Soc Rev* [Internet]. 2012;41(10):4030–66. Available from: https://doi.org/10.1039/C2CS15362A

69. Zhao H. What do we learn from enzyme behaviors in organic solvents? – Structural functionalization of ionic liquids for enzyme activation and stabilization. *Biotechnol Adv* [Internet]. 2020;45:107638. Available from: https://www.sciencedirect.com/science/article/pii/S0734975020301403

70. Fumino K, Reimann S, Ludwig R. Probing molecular interaction in ionic liquids by low frequency spectroscopy: Coulomb energy{,} hydrogen bonding and dispersion forces. *Phys Chem Chem Phys* [Internet]. 2014;16(40):21903–29. Available from: https://doi.org/10.1039/C4CP01476F

71. Hayyan M, Aissaoui T, Hashim MA, AlSaadi MA, Hayyan A. Triethylene glycol based deep eutectic solvents and their physical properties. *J Taiwan Inst Chem Eng* [Internet]. 2015;50:24–30. Available from: https://www.sciencedirect.com/science/article/pii/S1876107015000929

72. Wang J, Xu L, Jia G, Du J. Challenges and opportunities of ionic liquid electrolytes for rechargeable batteries. *Cryst Growth Des* [Internet]. 2022 Sep 7;22(9):5770–84. Available from: https://doi.org/10.1021/acs.cgd.2c00706

73. Barghamadi M, Best AS, Bhatt AI, Hollenkamp AF, Musameh M, Rees RJ, et al. Lithium–sulfur batteries—The solution is in the electrolyte{,} but is the electrolyte a solution? *Energy Environ Sci* [Internet]. 2014;7(12):3902–20. Available from: https://doi.org/10.1039/C4EE02192D

74. Plylahan N, Kerner M, Lim D-H, Matic A, Johansson P. Ionic liquid and hybrid ionic liquid/organic electrolytes for high temperature lithium-ion battery application. *Electrochim Acta* [Internet]. 2016;216:24–34. Available from: https://www.sciencedirect.com/science/article/pii/S0013468616317200

75. Kalhoff J, Eshetu GG, Bresser D, Passerini S. Safer electrolytes for lithium-ion batteries: State of the art and perspectives. *ChemSusChem* [Internet]. 2015;8(13):2154–75. Available from: https://chemistry-europe.onlinelibrary.wiley.com/doi/abs/10.1002/cssc.201500284

76. Liang F, Yu J, Chen J, Wang D, Lin C, Zhu C, et al. A novel boron-based ionic liquid electrolyte for high voltage lithium-ion batteries with outstanding cycling stability. *Electrochim Acta* [Internet]. 2018;283:111–20. Available from: https://www.sciencedirect.com/science/article/pii/S0013468618314580

77. Pal U, Girard GMA, O'Dell LA, Roy B, Wang X, Armand M, et al. Improved Li-ion transport by DME chelation in a novel ionic liquid-based hybrid electrolyte for Li–S battery application. *J Phys Chem C* [Internet]. 2018 Jul 5;122(26):14373–82. Available from: https://doi.org/10.1021/acs.jpcc.8b03909

78. Hofmann A, Migeot M, Hanemann T. Investigation of binary mixtures containing 1-ethyl-3-methylimidazolium bis(trifluoromethanesulfonyl)azanide and ethylene carbonate. *J Chem Eng Data* [Internet]. 2016 Jan 14;61(1):114–23. Available from: https://doi.org/10.1021/acs.jced.5b00338

79. Li W, Zhang Z, Han B, Hu S, Xie Y, Yang G. Effect of water and organic solvents on the ionic dissociation of ionic liquids. *J Phys Chem B* [Internet]. 2007 Jun 1;111(23):6452–6. Available from: https://doi.org/10.1021/jp071051m

80. Nickerson SD, Nofen EM, Chen H, Ngan M, Shindel B, Yu H, et al. A Combined experimental and molecular dynamics study of iodide-based ionic liquid and water mixtures. *J Phys Chem B* [Internet]. 2015 Jul 16;119(28):8764–72. Available from: https://doi.org/10.1021/acs.jpcb.5b04020

81. Comminges C, Barhdadi R, Laurent M, Troupel M. Determination of viscosity, ionic conductivity, and diffusion coefficients in some binary systems: Ionic liquids + molecular solvents. *J Chem Eng Data* [Internet]. 2006 Mar 1;51(2):680–5. Available from: https://doi.org/10.1021/je0504515

82. Lv Y, Xiao Y, Ma L, Zhi C, Chen S. Recent advances in electrolytes for "beyond aqueous" zinc-ion batteries. *Adv Mater* [Internet]. 2022;34(4):2106409. Available from: https://onlinelibrary.wiley.com/doi/abs/10.1002/adma.202106409

83. Vila J, Ginés P, Rilo E, Cabeza O, Varela LM. Great increase of the electrical conductivity of ionic liquids in aqueous solutions. *Fluid Phase Equilib* [Internet]. 2006;247(1):32–9. Available from: https://www.sciencedirect.com/science/article/pii/S0378381206002676

84. Zhang J, Wang Q, Cao Z. Effects of water on the structure and transport properties of room temperature ionic liquids and concentrated electrolyte solutions*. *Chin Phys B* [Internet]. 2020;29(8):87804. Available from: https://doi.org/10.1088/1674-1056/ab9c07

85. Annat G, Forsyth M, MacFarlane DR. Ionic liquid mixtures—Variations in physical properties and their origins in molecular structure. *J Phys Chem B* [Internet]. 2012 Jul 19;116(28):8251–8. Available from: https://doi.org/10.1021/jp3012602

86. Mohd Faridz Hilmy NI, Yahya WZN, Kurnia KA. Eutectic ionic liquids as potential electrolytes in dye-sensitized solar cells: Physicochemical and conductivity studies. *J Mol Liq* [Internet]. 2020;320:114381. Available from: https://www.sciencedirect.com/science/article/pii/S0167732220350133

87. Fox ET, Weaver JEF, Henderson WA. Tuning binary ionic liquid mixtures: Linking alkyl chain length to phase behavior and ionic conductivity. *J Phys Chem C* [Internet]. 2012 Mar 1;116(8):5270–4. Available from: https://doi.org/10.1021/jp300667h

88. Chatel G, Pereira JFB, Debbeti V, Wang H, Rogers RD. Mixing ionic liquids – "simple mixtures" or "double salts"? *Green Chem* [Internet]. 2014;16(4):2051–83. Available from: https://doi.org/10.1039/C3GC41389F

89. Wang Y-L, Li B, Sarman S, Mocci F, Lu Z-Y, Yuan J, et al. Microstructural and dynamical heterogeneities in ionic liquids. *Chem Rev* [Internet]. 2020 Jul 8;120(13):5798–877. Available from: https://doi.org/10.1021/acs.chemrev.9b00693

90. Huang J, Liu J, He J, Wu M, Qi S, Wang H, et al. Optimizing electrode/electrolyte interphases and Li-ion flux/solvation for lithium-metal batteries with qua-functional heptafluorobutyric anhydride. *Angew Chemie* [Internet]. 2021 Sep 13;133(38):20885–90. Available from: https://onlinelibrary.wiley.com/doi/10.1002/ange.202107957

91. von Aspern N, Leissing M, Wölke C, Diddens D, Kobayashi T, Börner M, et al. Non-flammable fluorinated phosphorus(III)-based electrolytes for advanced lithium-ion battery performance. *ChemElectroChem* [Internet]. 2020;7(6):1499–508. Available from: https://chemistry-europe.onlinelibrary. wiley.com/doi/abs/10.1002/celc.202000386

92. Curzons AD, Constable DC, Cunningham VL. Solvent selection guide: A guide to the integration of environmental, health and safety criteria into the selection of solvents. *Clean Prod Process* [Internet]. 1999;1(2):82–90. Available from: https://doi.org/10.1007/s100980050014

93. Gond R, van Ekeren W, Mogensen R, Naylor AJ, Younesi R. Non-flammable liquid electrolytes for safe batteries. *Mater Horiz* [Internet]. 2021;8(11):2913–28. Available from: https://doi.org/10.1039/D1MH00748C

94. Petkovic M, Seddon KR, Rebelo LPN, Silva Pereira C. Ionic liquids: A pathway to environmental acceptability. *Chem Soc Rev* [Internet]. 2011;40(3):1383–403. Available from: https://doi.org/10.1039/C004968A

95. Wei P, Pan X, Chen C-Y, Li H-Y, Yan X, Li C, et al. Emerging impacts of ionic liquids on eco-environmental safety and human health. *Chem Soc Rev* [Internet]. 2021;50(24):13609–27. Available from: https://doi.org/10.1039/D1CS00946J

96. Aspern N, Röschenthaler G -V., Winter M, Cekic-Laskovic I. Fluorine and lithium: Ideal partners for high-performance rechargeable battery electrolytes. *Angew Chemie Int Ed* [Internet]. 2019 Nov 4;58(45):15978–6000. Available from: https://onlinelibrary.wiley.com/doi/10.1002/anie.201901381

97. Haregewoin AM, Wotango AS, Hwang B-J. Electrolyte additives for lithium ion battery electrodes: Progress and perspectives. *Energy Environ Sci* [Internet]. 2016;9(6):1955–88. Available from: https://doi.org/10.1039/C6EE00123H

98. Wang AA, Greenbank S, Li G, Howey DA, Monroe CW. Current-driven solvent segregation in lithium-ion electrolytes. *Cell Reports Phys Sci* [Internet]. 2022 Sep;3(9):101047. Available from: https://linkinghub.elsevier.com/retrieve/pii/S2666386422003411

99. Moon J, Kim DO, Bekaert L, Song M, Chung J, Lee D, et al. Non-fluorinated non-solvating cosolvent enabling superior performance of lithium metal negative electrode battery. *Nat Commun* [Internet]. 2022 Aug 4;13(1):4538. Available from: https://www.nature.com/articles/s41467-022-32192-5

100. S. G. Guidance on Storage and Handling of Chlorinated Solventso [Internet]. 5th ed. Brussel; 2016. Available from: https://www.chlorinated-solvents.eu/wp-content/uploads/2019/12/EN_ECSA-Guidance-on-Storage-and-handling-of-chlorinated-solvents-January-2018.pdf

101. Gaida B, Brzęczek-Szafran A. Insights into the properties and potential applications of renewable carbohydrate-based ionic liquids: A review. *Molecules* [Internet]. 2020 Jul 20;25(14):3285. Available from: https://www.mdpi.com/1420-3049/25/14/3285

102. Mai NL, Ahn K, Koo Y-M. Methods for recovery of ionic liquids—A review. *Process Biochem* [Internet]. 2014;49(5):872–81. Available from: https://www.sciencedirect.com/science/article/pii/S1359511314000579

103. Ståhlberg T, Fu W, Woodley JM, Riisager A. Synthesis of 5-(hydroxymethyl) furfural in ionic liquids: Paving the way to renewable chemicals. *ChemSusChem* [Internet]. 2011;4(4):451–8. Available from: https://chemistry-europe.onlinelibrary. wiley.com/doi/abs/10.1002/cssc.201000374

104. Morina R, Merli D, Mustarelli P, Ferrara C. Lithium and cobalt recovery from lithium-ion battery waste via functional ionic liquid extraction for effective battery recycling. *ChemElectroChem* [Internet]. 2023;10(1):e202201059. Available from: https://chemistry-europe.onlinelibrary.wiley.com/doi/abs/10.1002/celc.202201059

105. Livi S, Baudoux J, Gérard J-F, Duchet-Rumeau J. Ionic liquids: A versatile platform for the design of a multifunctional epoxy networks 2.0 generation. *Prog Polym Sci* [Internet]. 2022;132:101581. Available from: https://www.sciencedirect.com/science/article/pii/S007967002200079X

106. Abbott AP, Ryder KS, König U. Electrofinishing of metals using eutectic based ionic liquids. *Trans IMF* [Internet]. 2008 Jul 1;86(4):196–204. Available from: https://doi.org/10.1179/174591908X327590

107. Kohli R. Chapter 1 - Removal of surface contaminants using ionic liquids. In: Kohli R, Mittal KLBT-D in SC and C, editors. Oxford: William Andrew Publishing; 2013. p. 1–63. Available from: https://www.sciencedirect.com/science/article/pii/B9781437778793000017

108. Patel KK, Singhal T, Pandey V, Sumangala TP, Sreekanth MS. Evolution and recent developments of high performance electrode material for supercapacitors: A review. *J Energy Storage* [Internet]. 2021;44:103366. Available from: https://www.sciencedirect.com/science/article/pii/S2352152X21010574

109. Cristina R. Ionic liquids synthesis – Methodologies. *Org Chem Curr Res*. 2014 Jan 1;04:e139.

110. Gujjala LKS, Kundu D, Dutta D, Kumar A, Bal M, Kumar A, et al. Advances in ionic liquids: Synthesis, environmental remediation and reusability. *J Mol Liq* [Internet]. 2024;396:123896. Available from: https://www.sciencedirect.com/science/article/pii/S0167732223027034

111. Minami I. Ionic liquids in tribology. *Molecules* [Internet]. 2009 Jun 24;14(6):2286–305. Available from: http://www.mdpi.com/1420-3049/14/6/2286

112. Hagberg J, Maples HA, Alvim KSP, Xu J, Johannisson W, Bismarck A, et al. Lithium iron phosphate coated carbon fiber electrodes for structural lithium ion batteries. *Compos Sci Technol* [Internet]. 2018;162:235–43. Available from: https://www.sciencedirect.com/science/article/pii/S0266353818302082

113. Zhou G, Li L, Wang D-W, Shan X, Pei S, Li F, et al. A flexible sulfur-graphene-polypropylene separator integrated electrode for advanced Li–S batteries. *Adv Mater* [Internet]. 2015 Jan 1;27(4):641–7. Available from: https://doi.org/10.1002/adma.201404210

114. Kim SW, Ryou M-H, Lee YM, Cho KY. Effect of liquid oil additive on lithium-ion battery ceramic composite separator prepared with an aqueous coating solution. *J Alloys Compd* [Internet]. 2016;675:341–7. Available from: https://www.sciencedirect.com/science/article/pii/S0925838816307162

115. Liu J, Li K, Zhang Q, Zhang X, Liang X, Yan J, et al. 3D tungsten disulfide/carbon nanotube networks as separator coatings and cathode additives for stable and fast lithium–sulfur batteries. *ACS Appl Mater Interfaces* [Internet]. 2021 Sep 29;13(38):45547–57. Available from: https://doi.org/10.1021/acsami.1c13193

116. Song Y-Z, Zhang Y, Yuan J-J, Lin C-E, Yin X, Sun C-C, et al. Fast assemble of polyphenol derived coatings on polypropylene separator for high performance lithium-ion batteries. *J Electroanal Chem* [Internet]. 2018;808:252–8. Available from: https://www.sciencedirect.com/science/article/pii/S1572665717308962

117. Kulacki KJ, Lamberti GA. Toxicity of imidazolium ionic liquids to freshwater algae. *Green Chem* [Internet]. 2008;10(1):104–10. Available from: https://doi.org/10.1039/B709289J

118. Magina S, Barros-Timmons A, Ventura SPM, Evtuguin DV. Evaluating the hazardous impact of ionic liquids – Challenges and opportunities. *J Hazard Mater* [Internet]. 2021;412:125215. Available from: https://www.sciencedirect.com/science/article/pii/S0304389421001783

119. Chatel G, Naffrechoux E, Draye M. Avoid the PCB mistakes: A more sustainable future for ionic liquids. *J Hazard Mater* [Internet]. 2017;324:773–80. Available from: https://www.sciencedirect.com/science/article/pii/S0304389416310901

120. No Title [Internet]. Available from: http://hdl.handle.net/2445/32193

121. Zhao D, Liao Y, Zhang Z. Toxicity of ionic liquids. *CLEAN – Soil, Air, Water* [Internet]. 2007 Feb 1;35(1):42–8. Available from: https://doi.org/10.1002/clen.200600015

122. Costa SPF, Azevedo AMO, Pinto PCAG, Saraiva MLMFS. Environmental impact of ionic liquids: Recent advances in (eco)toxicology and (bio)degradability. *ChemSusChem* [Internet]. 2017;10(11):2321–47. Available from: https://chemistry-europe.onlinelibrary.wiley.com/doi/abs/10.1002/cssc.201700261

123. Zhang Y, Zuo T-T, Popovic J, Lim K, Yin Y-X, Maier J, et al. Towards better Li metal anodes: Challenges and strategies. *Mater Today* [Internet]. 2020;33:56–74. Available from: https://www.sciencedirect.com/science/article/pii/S1369702119307837

124. Chen J, Li Q, Pollard TP, Fan X, Borodin O, Wang C. Electrolyte design for Li metal-free Li batteries. *Mater Today* [Internet]. 2020;39:118–26. Available from: https://www.sciencedirect.com/science/article/pii/S1369702120301061

125. Qin K, Holguin K, Mohammadiroudbari M, Huang J, Kim EYS, Hall R, et al. Strategies in structure and electrolyte design for high-performance lithium metal batteries. *Adv Funct Mater* [Internet]. 2021;31(15):2009694. Available from: https://onlinelibrary.wiley.com/doi/abs/10.1002/adfm.202009694

126. Ye H, Zhang Y, Yin Y-X, Cao F-F, Guo Y-G. An outlook on low-volume-change lithium metal anodes for long-life batteries. *ACS Cent Sci* [Internet]. 2020 May 27;6(5):661–71. Available from: https://doi.org/10.1021/acscentsci.0c00351

127. Torabifard H, Reed L, Berry MT, Hein JE, Menke E, Cisneros GA. Computational and experimental characterization of a pyrrolidinium-based ionic liquid for electrolyte applications. *J Chem Phys* [Internet]. 2017 Oct 4;147(16):161731. Available from: https://doi.org/10.1063/1.5004680

128. Isikli S, Ryan KM. Recent advances in solid-state polymer electrolytes and innovative ionic liquids based polymer electrolyte systems. *Curr Opin Electrochem* [Internet]. 2020 Jun;21:188–91. Available from: https://linkinghub.elsevier.com/retrieve/pii/S2451910320300223

129. Guerfi A, Duchesne S, Kobayashi Y, Vijh A, Zaghib K. LiFePO4 and graphite electrodes with ionic liquids based on bis(fluorosulfonyl)imide (FSI)– for Li-ion batteries. *J Power Sources* [Internet]. 2008;175(2):866–73. Available from: https://www.sciencedirect.com/science/article/pii/S0378775307018836

130. Wang D, Takiyama M, Hwang J, Matsumoto K, Hagiwara R. A Hexafluorophosphate-based ionic liquid as multifunctional interfacial layer between high voltage positive electrode and solid-state electrolyte for sodium secondary batteries. *Adv Energy Mater* [Internet]. 2023;13(30):2301020. Available from: https://onlinelibrary.wiley.com/doi/abs/10.1002/aenm.202301020

131. Zhao C-Z, Zhao B-C, Yan C, Zhang X-Q, Huang J-Q, Mo Y, et al. Liquid phase therapy to solid electrolyte–electrode interface in solid-state Li metal batteries: A review. *Energy Storage Mater* [Internet]. 2020;24:75–84. Available from: https://www.sciencedirect.com/science/article/pii/S2405829719308864

132. Xie Z, Wu Z, An X, Yoshida A, Wang Z, Hao X, et al. Bifunctional ionic liquid and conducting ceramic co-assisted solid polymer electrolyte membrane for quasi-solid-state lithium metal batteries. *J Memb Sci* [Internet]. 2019;586:122–9. Available from: https://www.sciencedirect.com/science/article/pii/S0376738819305459

133. Bao W, Hu Z, Wang Y, Jiang J, Huo S, Fan W, et al. Poly(ionic liquid)-functionalized graphene oxide towards ambient temperature operation of all-solid-state PEO-based polymer electrolyte lithium metal batteries. *Chem Eng J* [Internet]. 2022;437:135420. Available from: https://www.sciencedirect.com/science/article/pii/S1385894722009238

134. Jeon H, Hoang HA, Kim D. Flexible PVA/BMIMOTf/LLZTO composite electrolyte with liquid-comparable ionic conductivity for solid-state lithium metal battery. *J Energy Chem* [Internet]. 2022;74:128–39. Available from: https://www.sciencedirect.com/science/article/pii/S2095495622003746

135. Yuan Y, Peng X, Wang B, Xue K, Li Z, Ma Y, et al. Solvate ionic liquid-derived solid polymer electrolyte with lithium bis(oxalato) borate as a functional additive for solid-state lithium metal batteries. *J Mater Chem A* [Internet]. 2023;11(3):1301–11. Available from: https://doi.org/10.1039/D2TA07393E

136. Ortiz-Martínez VM, Gómez-Coma L, Pérez G, Ortiz A, Ortiz I. The roles of ionic liquids as new electrolytes in redox flow batteries. *Sep Purif Technol* [Internet]. 2020;252:117436. Available from: https://www.sciencedirect.com/science/article/pii/S1383586620319109

137. He X, Zhuang T, Ruan S, Xia X, Xia Y, Zhang J, et al. An innovative poly(ionic liquid) hydrogel-based anti-freezing electrolyte with high conductivity for supercapacitor. *Chem Eng J* [Internet]. 2023;466:143209. Available from: https://www.sciencedirect.com/science/article/pii/S138589472301940X

138. Zhou D, Liu R, Zhang J, Qi X, He Y-B, Li B, et al. In situ synthesis of hierarchical poly(ionic liquid)-based solid electrolytes for high-safety lithium-ion and sodium-ion batteries. *Nano Energy* [Internet]. 2017;33:45–54. Available from: https://www.sciencedirect.com/science/article/pii/S2211285517300320

139. Chen R, Hempelmann R. Ionic liquid-mediated aqueous redox flow batteries for high voltage applications. *Electrochem commun* [Internet]. 2016;70:56–9. Available from: https://www.sciencedirect.com/science/article/pii/S1388248116301564

140. Joseph A, Sobczak J, Żyła G, Mathew S. Ionic liquid and ionanofluid-based redox flow batteries—A mini review. *Energies* [Internet]. 2022;15(13). Available from: https://www.mdpi.com/1996-1073/15/13/4545

141. Chakrabarti MH, Mjalli FS, AlNashef IM, Hashim MA, Hussain MA, Bahadori L, et al. Prospects of applying ionic liquids and deep eutectic solvents for renewable energy storage by means of redox flow batteries. *Renew Sustain Energy Rev* [Internet]. 2014;30:254–70. Available from: https://www.sciencedirect.com/science/article/pii/S136403211300703X

142. Takechi K, Kato Y, Hase Y. A highly concentrated catholyte based on a solvate ionic liquid for rechargeable flow batteries. *Adv Mater* [Internet]. 2015;27(15):2501–6. Available from: https://onlinelibrary.wiley.com/doi/abs/ 10.1002/adma.201405840

143. Ejigu A, Greatorex-Davies PA, Walsh DA. Room temperature ionic liquid electrolytes for redox flow batteries. *Electrochem Commun* [Internet]. 2015;54:55–9. Available from: https://www.sciencedirect.com/science/article/ pii/S1388248115000260

144. Song J, Xu T, Gordin ML, Zhu P, Lv D, Jiang Y-B, et al. Nitrogen-doped mesoporous carbon promoted chemical adsorption of sulfur and fabrication of high-areal-capacity sulfur cathode with exceptional cycling stability for lithium-sulfur batteries. *Adv Funct Mater* [Internet]. 2014;24(9):1243– 50. Available from: https://onlinelibrary.wiley.com/doi/abs/10.1002/adfm. 201302631

145. Peng H-J, Hou T-Z, Zhang Q, Huang J-Q, Cheng X-B, Guo M-Q, et al. Strongly coupled interfaces between a heterogeneous carbon host and a sulfur-containing guest for highly stable lithium-sulfur batteries: Mechanistic insight into capacity degradation. *Adv Mater Interfaces* [Internet]. 2014 Oct 1;1(7):1400227. Available from: https://doi.org/10.1002/admi.201400227

146. Zhang B, Qin X, Li GR, Gao XP. Enhancement of long stability of sulfur cathode by encapsulating sulfur into micropores of carbon spheres. *Energy Environ Sci* [Internet]. 2010;3(10):1531–7. Available from: https://doi.org/10.1039/ C002639E

147. Sheng O, Jin C, Luo J, Yuan H, Fang C, Huang H, et al. Ionic conductivity promotion of polymer electrolyte with ionic liquid grafted oxides for all-solid-state lithium–sulfur batteries. *J Mater Chem A* [Internet]. 2017;5(25):12934–42. Available from: https://doi.org/10.1039/C7TA03699J

148. Josef E, Yan Y, Stan MC, Wellmann J, Vizintin A, Winter M, et al. Ionic liquids and their polymers in lithium-sulfur batteries. *Isr J Chem* [Internet]. 2019;59(9):832–42. Available from: https://onlinelibrary.wiley.com/doi/abs/ 10.1002/ijch.201800159

149. Yim T, Kwon M-S, Mun J, Lee KT. Room temperature ionic liquid-based electrolytes as an alternative to carbonate-based electrolytes. *Isr J Chem* [Internet]. 2015;55(5):586–98. Available from: https://onlinelibrary.wiley.com/doi/abs/ 10.1002/ijch.201400181

150. Hu Y, Pan J, Li Q, Ren Y, Qi H, Guo J, et al. Poly(ionic liquid)-based conductive interlayer as an efficient polysulfide adsorbent for a highly stable lithium–sulfur battery. *ACS Sustain Chem Eng* [Internet]. 2020 Aug 3;8(30):11396–403. Available from: https://doi.org/10.1021/acssuschemeng.0c03754

151. Barghamadi M, Best AS, Bhatt AI, Hollenkamp AF, Mahon PJ, Musameh M, et al. Effect of $LiNO_3$ additive and pyrrolidinium ionic liquid on the solid electrolyte interphase in the lithium–sulfur battery. *J Power Sources* [Internet]. 2015;295:212–20. Available from: https://www.sciencedirect.com/science/ article/pii/S0378775315300410

152. Hu J, Chen K, Li C. Nanostructured Li-rich fluoride coated by ionic liquid as high ion-conductivity solid electrolyte additive to suppress dendrite growth at Li metal anode. *ACS Appl Mater Interfaces* [Internet]. 2018 Oct 10;10(40):34322– 31. Available from: https://doi.org/10.1021/acsami.8b12579

153. Wang L, Ye Y, Chen N, Huang Y, Li L, Wu F, et al. Development and challenges of functional electrolytes for high-performance lithium–sulfur batteries. *Adv Funct Mater* [Internet]. 2018;28(38):1800919. Available from: https://onlinelibrary.wiley.com/doi/abs/10.1002/adfm.201800919

154. Liu B, Soares P, Checkles C, Zhao Y, Yu G. Three-dimensional hierarchical ternary nanostructures for high-performance Li-ion battery anodes. *Nano Lett* [Internet]. 2013 Jul 10;13(7):3414–9. Available from: https://doi.org/10.1021/nl401880v

155. Fu K (Kelvin), Gong Y, Hitz GT, McOwen DW, Li Y, Xu S, et al. Three-dimensional bilayer garnet solid electrolyte based high energy density lithium metal–sulfur batteries. *Energy Environ Sci* [Internet]. 2017;10(7):1568–75. Available from: https://doi.org/10.1039/C7EE01004D

156. Golodnitsky D, Nathan M, Yufit V, Strauss E, Freedman K, Burstein L, et al. Progress in three-dimensional (3D) Li-ion microbatteries. *Solid State Ionics* [Internet]. 2006;177(26):2811–9. Available from: https://www.sciencedirect.com/science/article/pii/S0167273806001172

157. Yang Y, Yuan W, Zhang X, Yuan Y, Wang C, Ye Y, et al. Overview on the applications of three-dimensional printing for rechargeable lithium-ion batteries. *Appl Energy* [Internet]. 2020;257:114002. Available from: https://www.sciencedirect.com/science/article/pii/S0306261919316897

158. Boz B, Dev T, Salvadori A, Schaefer JL. Review—Electrolyte and electrode designs for enhanced ion transport properties to enable high performance lithium batteries. *J Electrochem Soc* [Internet]. 2021;168(9):90501. Available from: https://doi.org/10.1149/1945-7111/ac1cc3

159. Ma J, Ma X, Zhang H, Chen F, Guan X, Niu J, et al. In-situ generation of poly(ionic liquid) flexible quasi-solid electrolyte supported by polyhedral oligomeric silsesquioxane/polyvinylidene fluoride electrospun membrane for lithium metal battery. *J Memb Sci* [Internet]. 2022;659:120811. Available from: https://www.sciencedirect.com/science/article/pii/S0376738822005567

160. He Y, Qiao Y, Chang Z, Zhou H. The potential of electrolyte filled MOF membranes as ionic sieves in rechargeable batteries. *Energy Environ Sci* [Internet]. 2019;12(8):2327–44. Available from: https://doi.org/10.1039/C8EE03651A

161. Thorat IV, Stephenson DE, Zacharias NA, Zaghib K, Harb JN, Wheeler DR. Quantifying tortuosity in porous Li-ion battery materials. *J Power Sources* [Internet]. 2009;188(2):592–600. Available from: https://www.sciencedirect.com/science/article/pii/S0378775308023574

162. Venugopal G, Moore J, Howard J, Pendalwar S. Characterization of microporous separators for lithium-ion batteries. *J Power Sources* [Internet]. 1999;77(1):34–41. Available from: https://www.sciencedirect.com/science/article/pii/S0378775398001682

163. Jeong H-S, Choi E-S, Lee S-Y, Kim JH. Evaporation-induced, close-packed silica nanoparticle-embedded nonwoven composite separator membranes for high-voltage/high-rate lithium-ion batteries: Advantageous effect of highly percolated, electrolyte-philic microporous architecture. *J Memb Sci* [Internet]. 2012;415–416:513–9. Available from: https://www.sciencedirect.com/science/article/pii/S037673881200422X

164. Kim J-H, Kim J-H, Kim J-M, Lee Y-G, Lee S-Y. Superlattice crystals–mimic, flexible/functional ceramic membranes: Beyond polymeric battery separators. *Adv Energy Mater* [Internet]. 2015;5(24):1500954. Available from: https://onlinelibrary.wiley.com/doi/abs/10.1002/aenm.201500954

165. Ryu J, Han D-Y, Hong D, Park S. A polymeric separator membrane with chemoresistance and high Li-ion flux for high-energy-density lithium metal batteries. *Energy Storage Mater* [Internet]. 2022;45:941–51. Available from: https://www.sciencedirect.com/science/article/pii/S2405829721006280

166. Urgoiti-Rodriguez M, Vaquero-Vílchez S, Mirandona-Olaeta A, de Luis R, Goikolea E, Costa CM, et al. Exploring ionic liquid-laden metal-organic framework composite materials as hybrid electrolytes in metal (ion) batteries. *Front Chem* [Internet]. 2022;10. Available from: https://www.frontiersin.org/articles/10.3389/fchem.2022.995063

167. Schaefer JL, Lu Y, Moganty SS, Agarwal P, Jayaprakash N, Archer LA. Electrolytes for high-energy lithium batteries. *Appl Nanosci* [Internet]. 2012;2(2):91–109. Available from: https://doi.org/10.1007/s13204-011-0044-x

168. Huang W-H, Li X-M, Yang X-F, Zhang X-X, Wang H-H, Wang H. The recent progress and perspectives on metal- and covalent-organic framework based solid-state electrolytes for lithium-ion batteries. *Mater Chem Front* [Internet]. 2021;5(9):3593–613. Available from: https://doi.org/10.1039/D0QM00936A

169. Wu C, Zeng W. Gel electrolyte for Li metal battery. *Chem Asian J* [Internet]. 2022;17(23):e202200816. Available from: https://onlinelibrary.wiley.com/doi/abs/10.1002/asia.202200816

170. Xiang J, Zhang Y, Zhang B, Yuan L, Liu X, Cheng Z, et al. A flame-retardant polymer electrolyte for high performance lithium metal batteries with an expanded operation temperature. *Energy Environ Sci* [Internet]. 2021;14(6):3510–21. Available from: https://doi.org/10.1039/D1EE00049G

171. Wang X, Salari M, Jiang D, Chapman Varela J, Anasori B, Wesolowski DJ, et al. Electrode material–ionic liquid coupling for electrochemical energy storage. *Nat Rev Mater* [Internet]. 2020;5(11):787–808. Available from: https://doi.org/10.1038/s41578-020-0218-9

172. Rehman A, Zeng X. Interfacial composition, structure, and properties of ionic liquids and conductive polymers for the construction of chemical sensors and biosensors: A perspective. *Curr Opin Electrochem* [Internet]. 2020;23:47–56. Available from: https://www.sciencedirect.com/science/article/pii/S2451910320300673

173. Yin L, Li S, Liu X, Yan T. Ionic liquid electrolytes in electric double layer capacitors. *Sci China Mater* [Internet]. 2019;62(11):1537–55. Available from: https://doi.org/10.1007/s40843-019-9458-3

174. Daneshvar F, Aziz A, Abdelkader AM, Zhang T, Sue H-J, Welland ME. Porous SnO_2 –Cu x O nanocomposite thin film on carbon nanotubes as electrodes for high performance supercapacitors. *Nanotechnology* [Internet]. 2019 Jan 4;30(1):015401. Available from: https://iopscience.iop.org/article/10.1088/1361-6528/aae5c6

175. Jiang J, Li Y, Liu J, Huang X, Yuan C, Lou XW (David). Recent advances in metal oxide-based electrode architecture design for electrochemical energy storage. *Adv Mater* [Internet]. 2012 Oct 2;24(38):5166–80. Available from: https://doi.org/10.1002/adma.201202146

176. Lokhande CD, Dubal DP, Joo O-S. Metal oxide thin film based supercapacitors. *Curr Appl Phys* [Internet]. 2011;11(3):255–70. Available from: https://www.sciencedirect.com/science/article/pii/S1567173910004773

177. Pervez A, Kim G, Bhaghavathi Parambath V, Cambaz M, Kuenzel M, Hekmatfar M, et al. Overcoming the interfacial limitations imposed by the solid-solid interface in solid-state batteries using ionic liquid-based interlayers. *Small*. 2020 Feb 27;16:2000279.

178. Morris MA, An H, Lutkenhaus JL, Epps THIII. Harnessing the power of plastics: Nanostructured polymer systems in lithium-ion batteries. *ACS Energy Lett* [Internet]. 2017 Aug 11;2(8):1919–36. Available from: https://doi.org/10.1021/acsenergylett.7b00368

179. Gu W, Yushin G. Review of nanostructured carbon materials for electrochemical capacitor applications: advantages and limitations of activated carbon, carbide-derived carbon, zeolite-templated carbon, carbon aerogels, carbon nanotubes, onion-like carbon, and graphene. *WIREs Energy Environ* [Internet]. 2014 Sep 1;3(5):424–73. Available from: https://doi.org/10.1002/wene.102

180. Ma D, Wang J. Inorganic electrochromic materials based on tungsten oxide and nickel oxide nanostructures. *Sci China Chem* [Internet]. 2017;60(1):54–62. Available from: https://doi.org/10.1007/s11426-016-0307-x

181. Chen S, Xing W, Duan J, Hu X, Qiao SZ. Nanostructured morphology control for efficient supercapacitor electrodes. *J Mater Chem A* [Internet]. 2013;1(9):2941–54. Available from: https://doi.org/10.1039/C2TA00627H

182. Vatamanu J, Hu Z, Bedrov D, Perez C, Gogotsi Y. Increasing energy storage in electrochemical capacitors with ionic liquid electrolytes and nanostructured carbon electrodes. *J Phys Chem Lett* [Internet]. 2013 Sep 5;4(17):2829–37. Available from: https://doi.org/10.1021/jz401472c

183. Simon P, Gogotsi Y. Capacitive energy storage in nanostructured carbon–electrolyte systems. *Acc Chem Res* [Internet]. 2013 May 21;46(5):1094–103. Available from: https://doi.org/10.1021/ar200306b

184. Liu X-M, Huang Z dong, Oh S woon, Zhang B, Ma P-C, Yuen MMF, et al. Carbon nanotube (CNT)-based composites as electrode material for rechargeable Li-ion batteries: A review. *Compos Sci Technol* [Internet]. 2012;72(2):121–44. Available from: https://www.sciencedirect.com/science/article/pii/S0266353811004118

185. Jung Y, Jeong YC, Kim JH, Kim YS, Kim T, Cho YS, et al. One step preparation and excellent performance of CNT yarn based flexible micro lithium ion batteries. *Energy Storage Mater* [Internet]. 2016;5:1–7. Available from: https://www.sciencedirect.com/science/article/pii/S240582971630085X

186. Zhou Q, Liu J, Gong X, Wang Z. A flexible and conductive connection introduced by cross-linked CNTs between submicron Si@C particles for better performance LIB anode. *Nanoscale Adv* [Internet]. 2021;3(8):2287–94. Available from: http://xlink.rsc.org/?DOI=D1NA00012H

187. Raccichini R, Varzi A, Passerini S, Scrosati B. The role of graphene for electrochemical energy storage. *Nat Mater* [Internet]. 2015;14(3):271–9. Available from: https://doi.org/10.1038/nmat4170

188. Wu HB, Chen JS, Hng HH, Wen (David) Lou X. Nanostructured metal oxide-based materials as advanced anodes for lithium-ion batteries. *Nanoscale* [Internet]. 2012;4(8):2526–42. Available from: https://doi.org/10.1039/C2NR11966H

189. Kaprans K, Mateuss J, Dorondo A, Bajars G, Kucinskis G, Lesnicenoks P, et al. Electrophoretically deposited α-Fe₂O₃ and TiO₂ composite anchored on rGO with excellent cycle performance as anode for lithium ion batteries. *Solid State Ionics* [Internet]. 2018;319:1–6. Available from: https://www.sciencedirect.com/science/article/pii/S0167273817307634

190. Kim U-H, Jun D-W, Park K-J, Zhang Q, Kaghazchi P, Aurbach D, et al. Pushing the limit of layered transition metal oxide cathodes for high-energy density rechargeable Li ion batteries. *Energy Environ Sci* [Internet]. 2018;11(5):1271–9. Available from: https://doi.org/10.1039/C8EE00227D

191. Ling JK, Jose R. Metal oxide composite cathode material for high energy density batteries. In: *Chemically Deposited Nanocrystalline Metal Oxide Thin Films: Synthesis, Characterizations, and Applications.* Ezema FI, Lokhande CD, Jose R, editors. Cham: Springer International Publishing; 2021. p. 509–30. Available from: https://doi.org/10.1007/978-3-030-68462-4_20

192. Wang L, Yue S, Zhang Q, Zhang Y, Li YR, Lewis CS, et al. Morphological and chemical tuning of high-energy-density metal oxides for lithium ion battery electrode applications. *ACS Energy Lett* [Internet]. 2017 Jun 9;2(6):1465–78. Available from: https://doi.org/10.1021/acsenergylett.7b00222

193. Mustaqeem M, Naikoo GA, Yarmohammadi M, Pedram MZ, Pourfarzad H, Dar RA, et al. Rational design of metal oxide based electrode materials for high performance supercapacitors – A review. *J Energy Storage* [Internet]. 2022;55:105419. Available from: https://www.sciencedirect.com/science/article/pii/S2352152X22014116

194. Dai M, Zhao D, Wu X. Research progress on transition metal oxide based electrode materials for asymmetric hybrid capacitors. *Chin Chem Lett* [Internet]. 2020;31(9):2177–88. Available from: https://www.sciencedirect.com/science/article/pii/S1001841720300772

195. Ma M, Du M, Liu Y, Lü H, Yang J, Hao Z, et al. Electrode particulate materials for advanced rechargeable batteries: A review. *Particuology* [Internet]. 2024;86:160–81. Available from: https://www.sciencedirect.com/science/article/pii/S1674200123001141

196. Stephenson T, Li Z, Olsen B, Mitlin D. Lithium ion battery applications of molybdenum disulfide (MoS₂) nanocomposites. *Energy Environ Sci* [Internet]. 2014;7(1):209–31. Available from: https://doi.org/10.1039/C3EE42591F

197. Jena KK, Mayyas AT, Mohanty B, Jena BK, Jos JR, AlFantazi A, et al. Recycling of electrode materials from spent lithium-ion batteries to develop graphene nanosheets and graphene–molybdenum disulfide nanohybrid: Environmental benefits, analysis of supercapacitor performance, and influence of density functional theory calculations. *Energy & Fuels* [Internet]. 2022 Feb 17;36(4):2159–70. Available from: https://doi.org/10.1021/acs.energyfuels.1c03789

198. Zhang X, Wang H, Wang G. Cobalt sulfide nanoparticles anchored in three-dimensional carbon nanosheet networks for lithium and sodium ion batteries with enhanced electrochemical performance. *J Colloid Interface Sci* [Internet]. 2017;492:41–50. Available from: https://www.sciencedirect.com/science/article/pii/S0021979716310797

199. Xiao Y, Lee SH, Sun Y-K. The application of metal sulfides in sodium ion batteries. *Adv Energy Mater* [Internet]. 2017 Feb 1;7(3):1601329. Available from: https://doi.org/10.1002/aenm.201601329

200. Zhao J, Zhang Y, Wang Y, Li H, Peng Y. The application of nanostructured transition metal sulfides as anodes for lithium ion batteries. *J Energy Chem* [Internet]. 2018;27(6):1536–54. Available from: https://www.sciencedirect.com/science/article/pii/S2095495617311762

201. Li Y, Wu F, Qian J, Zhang M, Yuan Y, Bai Y, et al. Metal chalcogenides with heterostructures for high-performance rechargeable batteries. *Small Sci* [Internet]. 2021;1(9):2100012. Available from: https://onlinelibrary.wiley.com/doi/abs/10.1002/smsc.202100012

202. Ji L, Lin Z, Alcoutlabi M, Zhang X. Recent developments in nanostructured anode materials for rechargeable lithium-ion batteries. *Energy Environ Sci* [Internet]. 2011;4(8):2682–99. Available from: https://doi.org/10.1039/C0EE00699H

203. Rhodes K, Dudney N, Lara-Curzio E, Daniel C. Understanding the degradation of silicon electrodes for lithium-ion batteries using acoustic emission. *J Electrochem Soc* [Internet]. 2010;157(12):A1354. Available from: https://doi.org/10.1149/1.3489374

204. Luo W, Chen X, Xia Y, Chen M, Wang L, Wang Q, et al. Surface and interface engineering of silicon-based anode materials for lithium-ion batteries. *Adv Energy Mater* [Internet]. 2017;7(24):1701083. Available from: https://onlinelibrary.wiley.com/doi/abs/10.1002/aenm.201701083

205. Yuca N, Taskin OS, Arici E. An overview on efforts to enhance the Si electrode stability for lithium ion batteries. *Energy Storage* [Internet]. 2020;2(1):e94. Available from: https://onlinelibrary.wiley.com/doi/abs/10.1002/est2.94

206. Lin D, Lu Z, Hsu P-C, Lee HR, Liu N, Zhao J, et al. A high tap density secondary silicon particle anode fabricated by scalable mechanical pressing for lithium-ion batteries. *Energy Environ Sci* [Internet]. 2015;8(8):2371–6. Available from: https://doi.org/10.1039/C5EE01363A

207. Martha SK, Elias L, Ghosh S. Nanostructured 3D (three dimensional) electrode architectures of silicon for high-performance Li-ion batteries. In: *Silicon Anode Systems for Lithium-Ion Batteries* [Internet]. Kumta PN, Hepp AF, Datta MK, Velikokhatnyi OI, editors. Elsevier; 2022. p. 331–71. Available from: https://www.sciencedirect.com/science/article/pii/B978012819660100013X

208. Saddique J, Wu M, Ali W, Xu X, Jiang Z-G, Tong L, et al. Opportunities and challenges of nano Si/C composites in lithium ion battery: A mini review. *J Alloys Compd* [Internet]. 2024;978:173507. Available from: https://www.sciencedirect.com/science/article/pii/S0925838824000938

209. Su Y-Z, Fu Y-C, Wei Y-M, Yan J-W, Mao B-W. The electrode/ionic liquid interface: Electric double layer and metal electrodeposition. *ChemPhysChem* [Internet]. 2010 Sep 10;11(13):2764–78. Available from: https://doi.org/10.1002/cphc.201000278

210. Arano K, Begic S, Chen F, Rakov D, Mazouzi D, Gautier N, et al. Tuning the formation and structure of the silicon electrode/ionic liquid electrolyte interphase in superconcentrated ionic liquids. *ACS Appl Mater Interfaces* [Internet]. 2021 Jun 23;13(24):28281–94. Available from: https://doi.org/10.1021/acsami.1c06465

211. Yamaguchi K, Domi Y, Usui H, Shimizu M, Morishita S, Yodoya S, et al. Effect of film-forming additive in ionic liquid electrolyte on electrochemical performance of Si negative-electrode for LIBs. *J Electrochem Soc* [Internet]. 2019;166(2):A268. Available from: https://doi.org/10.1149/2.0971902jes

212. Lashkari S, Pal R, Pope MA. ionic liquid/non-ionic surfactant mixtures as versatile, non-volatile electrolytes: Double-layer capacitance and conductivity. *J Electrochem Soc* [Internet]. 2022;169(4):40513. Available from: https://doi.org/10.1149/1945-7111/ac62c7

213. Hoffmann V, Pulletikurthi G, Carstens T, Lahiri A, Borodin A, Schammer M, et al. Influence of a silver salt on the nanostructure of a Au(111)/ionic liquid interface: an atomic force microscopy study and theoretical concepts. *Phys Chem Chem Phys* [Internet]. 2018;20(7):4760–71. Available from: https://doi.org/10.1039/C7CP08243F

214. Das D, Manna S, Puravankara S. Electrolytes, additives and binders for NMC cathodes in Li-ion batteries—A review. *Batteries* [Internet]. 2023;9(4). Available from: https://www.mdpi.com/2313-0105/9/4/193

215. Wang L, Wu F, Yao Y, Zhang C. Lithium metal anode in electrochemical perspective. *ChemElectroChem* [Internet]. 2024 Mar 27;n/a(n/a):e202400019. Available from: https://doi.org/10.1002/celc.202400019

216. Vatamanu J, Cao L, Borodin O, Bedrov D, Smith GD. On the influence of surface topography on the electric double layer structure and differential capacitance of graphite/ionic liquid interfaces. *J Phys Chem Lett* [Internet]. 2011 Sep 1;2(17):2267–72. Available from: https://doi.org/10.1021/jz200879a

217. Rakov D, Hasanpoor M, Baskin A, Lawson JW, Chen F, Cherepanov PV, et al. Stable and efficient lithium metal anode cycling through understanding the effects of electrolyte composition and electrode preconditioning. *Chem Mater* [Internet]. 2022 Jan 11;34(1):165–77. Available from: https://doi.org/10.1021/acs.chemmater.1c02981

218. Haskins JB, Wu JJ, Lawson JW. Computational and experimental study of Li-doped ionic liquids at electrified interfaces. *J Phys Chem C* [Internet]. 2016 Jun 9;120(22):11993–2011. Available from: https://doi.org/10.1021/acs.jpcc.6b02449

219. Smith AM, Perkin S. Influence of lithium solutes on double-layer structure of ionic liquids. *J Phys Chem Lett* [Internet]. 2015 Dec 3;6(23):4857–61. Available from: https://doi.org/10.1021/acs.jpclett.5b02166

220. Ma K, Forsman J, Woodward CE. Influence of ion pairing in ionic liquids on electrical double layer structures and surface force using classical density functional approach. *J Chem Phys* [Internet]. 2015 May 6;142(17):174704. Available from: https://doi.org/10.1063/1.4919314

221. Zhao Y, Stein P, Bai Y, Al-Siraj M, Yang Y, Xu B-X. A review on modeling of electro-chemo-mechanics in lithium-ion batteries. *J Power Sources* [Internet]. 2019;413:259–83. Available from: https://www.sciencedirect.com/science/article/pii/S0378775318313624

222. Zhu J, Wierzbicki T, Li W. A review of safety-focused mechanical modeling of commercial lithium-ion batteries. *J Power Sources* [Internet]. 2018;378:153–68. Available from: https://www.sciencedirect.com/science/article/pii/S0378775317316361

223. Weingärtner H. Understanding ionic liquids at the molecular level: Facts, problems, and controversies. *Angew Chemie Int Ed* [Internet]. 2008;47(4):654–70. Available from: https://onlinelibrary.wiley.com/doi/abs/10.1002/anie.200604951

224. Lv C, Zhou X, Zhong L, Yan C, Srinivasan M, Seh ZW, et al. Machine learning: An advanced platform for materials development and state prediction in lithium-ion batteries. *Adv Mater* [Internet]. 2022;34(25):2101474. Available from: https://onlinelibrary.wiley.com/doi/abs/10.1002/adma.202101474

225. Schwarz K, Sundararaman R. The electrochemical interface in first-principles calculations. *Surf Sci Rep* [Internet]. 2020;75(2):100492. Available from: https://www.sciencedirect.com/science/article/pii/S0167572920300133

226. Bedrov D, Piquemal J-P, Borodin O, MacKerell ADJ, Roux B, Schröder C. Molecular dynamics simulations of ionic liquids and electrolytes using polarizable force fields. *Chem Rev* [Internet]. 2019 Jul 10;119(13):7940–95. Available from: https://doi.org/10.1021/acs.chemrev.8b00763

227. Tepermeister M, Bosnjak N, Dai J, Zhang X, Kielar SM, Wang Z, et al. Soft ionics: Governing physics and state of technologies. *Front Phys* [Internet]. 2022 Jul 11;10. Available from: https://www.frontiersin.org/articles/10.3389/fphy.2022.890845/full

228. Wang A, Kadam S, Li H, Shi S, Qi Y. Review on modeling of the anode solid electrolyte interphase (SEI) for lithium-ion batteries. *npj Comput Mater* [Internet]. 2018;4(1):15. Available from: https://doi.org/10.1038/s41524-018-0064-0

229. Xu B, Fell CR, Chi M, Meng YS. Identifying surface structural changes in layered Li-excess nickel manganese oxides in high voltage lithium ion batteries: A joint experimental and theoretical study. *Energy Environ Sci* [Internet]. 2011;4(6):2223–33. Available from: https://doi.org/10.1039/C1EE01131F

230. Edge JS, O'Kane S, Prosser R, Kirkaldy ND, Patel AN, Hales A, et al. Lithium ion battery degradation: What you need to know. *Phys Chem Chem Phys* [Internet]. 2021;23(14):8200–21. Available from: http://xlink.rsc.org/?DOI=D1CP00359C

231. Shi S, Gao J, Liu Y, Zhao Y, Wu Q, Ju W, et al. Multi-scale computation methods: Their applications in lithium-ion battery research and development*. *Chin Phys B* [Internet]. 2016;25(1):18212. Available from: https://doi.org/10.1088/1674-1056/25/1/018212

232. Liu Y, Zhou Q, Cui G. Machine learning boosting the development of advanced lithium batteries. *Small Methods* [Internet]. 2021 Aug 1;5(8):2100442. Available from: https://doi.org/10.1002/smtd.202100442

233. Xu Z, Ma M, Liu P. Self-energy-modified Poisson-Nernst-Planck equations: WKB approximation and finite-difference approaches. *Phys Rev E* [Internet]. 2014;90(1):13307. Available from: https://link.aps.org/doi/10.1103/PhysRevE.90.013307

234. Yao N, Chen X, Fu Z-H, Zhang Q. Applying classical, ab initio, and machine-learning molecular dynamics simulations to the liquid electrolyte for rechargeable batteries. *Chem Rev* [Internet]. 2022 Jun 22;122(12):10970–1021. Available from: https://doi.org/10.1021/acs.chemrev.1c00904

235. Ong MT, Verners O, Draeger EW, van Duin ACT, Lordi V, Pask JE. Lithium ion solvation and diffusion in bulk organic electrolytes from first-principles and classical reactive molecular dynamics. *J Phys Chem B* [Internet]. 2015 Jan 29;119(4):1535–45. Available from: https://doi.org/10.1021/jp508184f

236. Pham TA, Kweon KE, Samanta A, Lordi V, Pask JE. Solvation and dynamics of sodium and potassium in ethylene carbonate from ab initio molecular dynamics simulations. *J Phys Chem C* [Internet]. 2017 Oct 12;121(40):21913–20. Available from: https://doi.org/10.1021/acs.jpcc.7b06457

237. Atkin R, Borisenko N, Drüschler M, Endres F, Hayes R, Huber B, et al. Structure and dynamics of the interfacial layer between ionic liquids and electrode materials. *J Mol Liq* [Internet]. 2014;192:44–54. Available from: https://www.sciencedirect.com/science/article/pii/S0167732213002699

238. Feng G, Huang J, Sumpter BG, Meunier V, Qiao R. Structure and dynamics of electrical double layers in organic electrolytes. *Phys Chem Chem Phys* [Internet]. 2010;12(20):5468–79. Available from: https://doi.org/10.1039/C000451K

239. Boyer MJ, Vilčiauskas L, Hwang GS. Structure and Li+ ion transport in a mixed carbonate/LiPF6 electrolyte near graphite electrode surfaces: A molecular dynamics study. *Phys Chem Chem Phys* [Internet]. 2016;18(40):27868–76. Available from: https://doi.org/10.1039/C6CP05140E

240. Xiao W, Yang Q, Zhu S. Comparing ion transport in ionic liquids and polymerized ionic liquids. *Sci Rep* [Internet]. 2020;10(1):7825. Available from: https://doi.org/10.1038/s41598-020-64689-8

241. Borodin O, Ren X, Vatamanu J, von Wald Cresce A, Knap J, Xu K. Modeling insight into battery electrolyte electrochemical stability and interfacial structure. *Acc Chem Res*. 2017 Dec;50(12):2886–94.

242. Liu Y, Sun Z, Sun X, Lin Y, Tan K, Sun J, et al. Construction of hierarchical nanotubes assembled from ultrathin V3S4@C nanosheets towards alkali-ion batteries with ion-dependent electrochemical mechanisms. *Angew Chemie Int Ed*. 2020 Feb;59(6):2473–82.

243. Lian C, Zhao S, Liu H, Wu J. Time-dependent density functional theory for the charging kinetics of electric double layer containing room-temperature ionic liquids. *J Chem Phys*. 2016 Nov;145(20).

244. Melander MM, Kuisma MJ, Christensen TEK, Honkala K. Grand-canonical approach to density functional theory of electrocatalytic systems: Thermodynamics of solid-liquid interfaces at constant ion and electrode potentials. *J Chem Phys*. 2019 Jan;150(4).

245. Ma J, Zhao S, Li Z. New crowding states of ionic liquid induced by configuration change of ion adsorption on charged electrode. *Electrochim Acta*. 2022;425:140692.

246. Maftoon-Azad L. Electrochemical stability windows of Ali-cyclic ionic liquids as lithium metal battery Electrolytes: A computational approach. *J Mol Liq* [Internet]. 2021 Dec;343:117589. Available from: https://linkinghub.elsevier.com/retrieve/pii/S016773222102314X

247. Sadeghian H, Maftoon-azad L, Jalali T. How ionic structure governs bulk properties: charge lever moments of alicyclic ionic liquids utilized in lithium metal batteries. *J Electrochem Soc* [Internet]. 2022 Jul 1;169(7):070521. Available from: https://iopscience.iop.org/article/10.1149/1945-7111/ac7e70

248. Maftoon-Azad L, Nazari F. Anion-cation, anion-lithium, cation-lithium and ion pair-lithium interactions in alicyclic ammonium based ionic liquids as electrolytes of lithium metal batteries. *J Mol Liq* [Internet]. 2017 Sep;242:1228–35. Available from: https://linkinghub.elsevier.com/retrieve/pii/S0167732217313338

249. Lu H, Nordholm S, Woodward CE, Forsman J. Ionic liquid interface at an electrode: Simulations of electrochemical properties using an asymmetric restricted primitive model. *J Phys Condens Matter*. 2018 Feb;30(7):074004.

250. Tu Y-J, Delmerico S, McDaniel JG. Inner layer capacitance of organic electrolytes from constant voltage molecular dynamics. *J Phys Chem C*. 2020 Feb;124(5):2907–22.

251. Tian G. Electrical double-layer structure and property of ionic liquid-electrode system for electrochemical applications. In: *Nanotechnology-Based Industrial Applications of Ionic Liquids*. Inamuddin, Asiri AM, editors. Cham: Springer International Publishing; 2020. p. 177–220.

252. Damaskin BB, Petrii OA. Historical development of theories of the electrochemical double layer. *J Solid State Electrochem*. 2011;15(7):1317–34.

253. Yoo K, Deshpande A, Banerjee S, Dutta P. Electrochemical model for ionic liquid electrolytes in lithium batteries. *Electrochim Acta*. 2015;176:301–10.

254. Latz A, Zausch J. Multiscale modeling of lithium ion batteries: Thermal aspects. *Beilstein J Nanotechnol*. 2015;6:987–1007.

255. Jiang X, Huang J, Zhao H, Sumpter BG, Qiao R. Dynamics of electrical double layer formation in room-temperature ionic liquids under constant-current charging conditions. *J Phys Condens Matter*. 2014 Jul;26(28):284109.

256. Girotto M, Colla T, dos Santos AP, Levin Y. Lattice model of an ionic liquid at an electrified interface. *J Phys Chem B*. 2017 Jul;121(26):6408–15.

257. Li G, Monroe CW. Modeling lithium transport and electrodeposition in ionic-liquid based electrolytes. *Front Energy Res*. 2021 May;9:660081.

258. Gao X, Liu X, He R, Wang M, Xie W, Brandon NP, et al. Designed high-performance lithium-ion battery electrodes using a novel hybrid model-data driven approach. *Energy Storage Mater*. 2021;36:435–58.

259. Li X, Han B, Yang X, Deng Z, Zou Y, Shi X, et al. Three-dimensional visualization of lithium metal anode via low-dose cryogenic electron microscopy tomography. *iScience*. 2021 Dec;24(12):103418.

260. Steinrück H-P. Recent developments in the study of ionic liquid interfaces using X-ray photoelectron spectroscopy and potential future directions. *Phys Chem Chem Phys* [Internet]. 2012;14(15):5010–29. Available from: https://doi.org/10.1039/C2CP24087D

261. Yokota Y, Mino Y, Kanai Y, Utsunomiya T, Imanishi A, Fukui K. Electronic-state changes of ferrocene-terminated self-assembled monolayers induced by molecularly thin ionic liquid layers: A combined atomic force microscopy, X-ray photoelectron spectroscopy, and ultraviolet photoelectron spectroscopy study. *J Phys Chem C* [Internet]. 2015 Aug 13;119(32):18467–80. Available from: https://doi.org/10.1021/acs.jpcc.5b05682

262. Patra A, K. N, Jose JR, Sahoo S, Chakraborty B, Rout CS. Understanding the charge storage mechanism of supercapacitors: In situ/operando spectroscopic approaches and theoretical investigations. *J Mater Chem A* [Internet]. 2021;9(46):25852–91. Available from: https://doi.org/10.1039/D1TA07401F

263. Hayes R, Warr GG, Atkin R. Structure and nanostructure in ionic liquids. *Chem Rev* [Internet]. 2015 Jul 8;115(13):6357–426. Available from: https://doi.org/10.1021/cr500411q

264. Appiah WA, Stark A, Lysgaard S, Busk J, Jankowski P, Chang JH, et al. Unveiling the plating-stripping mechanism in aluminum batteries with imidazolium-based electrolytes: A hierarchical model based on experiments and ab initio simulations. *Chem Eng J* [Internet]. 2023;472:144995. Available from: https://www.sciencedirect.com/science/article/pii/S1385894723037269

265. Jayakody NK, Fraenza CC, Greenbaum SG, Ashby D, Dunn BS. NMR relaxometry and diffusometry analysis of dynamics in ionic liquids and ionogels for use in lithium-ion batteries. *J Phys Chem B* [Internet]. 2020 Aug 6;124(31):6843–56. Available from: https://doi.org/10.1021/acs.jpcb.0c02755

266. Oliver TA, Terejanu G, Simmons CS, Moser RD. Validating predictions of unobserved quantities. *Comput Methods Appl Mech Eng* [Internet]. 2015;283:1310–35. Available from: https://www.sciencedirect.com/science/article/pii/S004578251400293X

267. Karniadakis GE, Kevrekidis IG, Lu L, Perdikaris P, Wang S, Yang L. Physics-informed machine learning. *Nat Rev Phys* [Internet]. 2021;3(6):422–40. Available from: https://doi.org/10.1038/s42254-021-00314-5

268. Boomsma W, Ferkinghoff-Borg J, Lindorff-Larsen K. Combining experiments and simulations using the maximum entropy principle. *PLOS Comput Biol* [Internet]. 2014 Feb 20;10(2):e1003406. Available from: https://doi.org/10.1371/journal.pcbi.1003406

269. Graebner C. How to relate models to reality? An epistemological framework for the validation and verification of computational models. *J Artif Soc Soc Simul* [Internet]. 2018;21(3):8. Available from: http://jasss.soc.surrey.ac.uk/21/3/8.html

270. Uhrmacher AM, Brailsford S, Liu J, Rabe M, Tolk A. Panel — Reproducible research in discrete event simulation — A must or rather a maybe? In: *2016 Winter Simulation Conference (WSC)*. 2016. p. 1301–15.

271. Alarid-Escudero F, Krijkamp EM, Pechlivanoglou P, Jalal H, Kao S-YZ, Yang A, et al. A need for change! A coding framework for improving transparency in decision modeling. *Pharmacoeconomics* [Internet]. 2019;37(11):1329–39. Available from: https://doi.org/10.1007/s40273-019-00837-x

272. Mulugeta L, Drach A, Erdemir A, Hunt CA, Horner M, Ku JP, et al. Credibility, replicability, and reproducibility in simulation for biomedicine and clinical applications in neuroscience. *Front Neuroinform* [Internet]. 2018;12. Available from: https://www.frontiersin.org/articles/10.3389/fninf.2018.00018

273. Lessmann S, Baesens B, Mues C, Pietsch S. Benchmarking classification models for software defect prediction: A proposed framework and novel findings. *IEEE Trans Softw Eng*. 2008;34(4):485–96.

274. Yao Y-X, Yan C, Zhang Q. Emerging interfacial chemistry of graphite anodes in lithium-ion batteries. *Chem Commun* [Internet]. 2020;56(93):14570–84. Available from: https://doi.org/10.1039/D0CC05084A

275. Balbuena PB, Wang YX. *Lithium-Ion Batteries: Solid-Electrolyte Interphase*. World Scientific; 2004.

276. Li Y, Lu Y, Adelhelm P, Titirici M-M, Hu Y-S. Intercalation chemistry of graphite: Alkali metal ions and beyond. *Chem Soc Rev* [Internet]. 2019;48(17):4655–87. Available from: https://doi.org/10.1039/C9CS00162J

277. Zhang Z, Yao T, Wang E, Sun B, Sun K, Peng Z. Unlocking the low-temperature potential of propylene carbonate to −30 °C via N-methylpyrrolidone. *ACS Appl Mater Interfaces* [Internet]. 2022 Oct 12;14(40):45484–93. Available from: https://doi.org/10.1021/acsami.2c13667

278. Limaye AM. *Physical Models and Statistical Methods for Understanding Electrochemical Kinetics* [Internet]. MIT; 2022. Available from: http://rightsstatements.org/page/InC-EDU/1.0/

279. Ageed Z, Mahmood MR, Sadeeq M, Abdulrazzaq MB, Dino H. Cloud computing resources impacts on heavy-load parallel processing approaches. *IOSR J Comput Eng.* 2020;22(3):30–41.

280. D'Angelo G, Marzolla M. New trends in parallel and distributed simulation: From many-cores to cloud computing. *Simul Model Pract Theory.* 2014;49:320–35.

281. Li S, Marsaglia N, Garth C, Woodring J, Clyne J, Childs H. Data reduction techniques for simulation, visualization and data analysis. *Comput Graph Forum* [Internet]. 2018 Sep 1;37(6):422–47. Available from: https://doi.org/10.1111/cgf.13336

282. Li D, Kong Y, Zheng Z, Pan J. Recent advances in big data analytics. In: *The Palgrave Handbook of Operations Research.* Salhi S, Boylan J, editors. Cham: Springer International Publishing; 2022. p. 805–34. Available from: https://doi.org/10.1007/978-3-030-96935-6_25

283. Jones DC, Ruzzo WL, Peng X, Katze MG. Compression of next-generation sequencing reads aided by highly efficient de novo assembly. *Nucleic Acids Res* [Internet]. 2012 Dec 1;40(22):e171–e171. Available from: https://doi.org/10.1093/nar/gks754

284. Hach F, Numanagić I, Alkan C, Sahinalp SC. SCALCE: Boosting sequence compression algorithms using locally consistent encoding. *Bioinformatics* [Internet]. 2012 Dec 1;28(23):3051–7. Available from: https://doi.org/10.1093/bioinformatics/bts593

285. Rahman MA, Hamada M. Burrows–wheeler transform based lossless text compression using keys and Huffman coding. *Symmetry (Basel)* [Internet]. 2020;12(10). Available from: https://www.mdpi.com/2073-8994/12/10/1654

286. Hübbe N, Kunkel J. Reducing the HPC-datastorage footprint with MAFISC—multidimensional adaptive filtering improved scientific data compression. *Comput Sci - Res Dev* [Internet]. 2013;28(2):231–9. Available from: https://doi.org/10.1007/s00450-012-0222-4

287. Liu T, Wang J, Liu Q, Alibhai S, Lu T, He X. High-ratio lossy compression: Exploring the autoencoder to compress scientific data. IEEE Trans Big Data. 2023;9(1):22–36.

288. Sallee P. Model-Based Steganography. In: *Digital Watermarking.* Kalker T, Cox I, Ro YM, editors. Berlin, Heidelberg: Springer Berlin Heidelberg; 2004. p. 154–67.

289. Li Q. Overview of data visualization. In: *Embodying Data: Chinese Aesthetics, Interactive Visualization and Gaming Technologies* [Internet]. Singapore: Springer Singapore; 2020. p. 17–47. Available from: https://doi.org/10.1007/978-981-15-5069-0_2

290. Khan M, Khan SS. Data and information visualization methods, and interactive mechanisms: A survey. *Int J Comput Appl.* 2011;34(1):1–14.

291. Aristodemou L, Tietze F. The state-of-the-art on intellectual property analytics (IPA): A literature review on artificial intelligence, machine learning and deep learning methods for analysing intellectual property (IP) data. *World Pat Inf* [Internet]. 2018;55:37–51. Available from: https://www.sciencedirect.com/science/article/pii/S0172219018300103

292. Rani KS, Kumari M, Singh VB, Sharma M. Deep learning with big data: An emerging trend. In: *2019 19th International Conference on Computational Science and Its Applications (ICCSA).* 2019. p. 93–101.

293. Zhou T, Song Z, Sundmacher K. Big data creates new opportunities for materials research: A review on methods and applications of machine learning for materials design. *Engineering* [Internet]. 2019;5(6):1017–26. Available from: https://www.sciencedirect.com/science/article/pii/S2095809918313559

294. Koutsoukos S, Philippi F, Malaret F, Welton T. A review on machine learning algorithms for the ionic liquid chemical space. *Chem Sci* [Internet]. 2021;12(20):6820–43. Available from: https://doi.org/10.1039/D1SC01000J

295. Liu Y, Zhao T, Ju W, Shi S. Materials discovery and design using machine learning. *J Mater* [Internet]. 2017;3(3):159–77. Available from: https://www.sciencedirect.com/science/article/pii/S2352847817300515

296. Pedregosa F, Varoquaux G, Gramfort A, Michel V, Thirion B, Grisel O, et al. Scikit-learn: Machine Learning in {P}ython. *J Mach Learn Res*. 2011;12:2825–30.

297. Lara JD, Barrows C, Thom D, Krishnamurthy D, Callaway D. PowerSystems. jl — A power system data management package for large scale modeling. *SoftwareX* [Internet]. 2021;15:100747. Available from: https://www.sciencedirect.com/science/article/pii/S2352711021000765

298. Kanoje S, Powar V, Mukhopadhyay D. *Using MongoDB for Social Networking Website*. 2015.

299. Shah JK, Marin-Rimoldi E, Mullen RG, Keene BP, Khan S, Paluch AS, et al. Cassandra: An open source Monte Carlo package for molecular simulation. *J Comput Chem*. 2017;38(19):1727–39.

300. Khatri V, Brown CV. Designing data governance. *Commun ACM* [Internet]. 2010;53(1):148–152. Available from: https://doi.org/10.1145/1629175.1629210

301. No Title [Internet]. Available from: https://www.grapestechsolutions.com/blog/advantages-of-data-management-services/w.grapestechsolutions.com/blog/advantages-of-data-management-services/

302. Van Noorden R. Online collaboration: Scientists and the social network. *Nature* [Internet]. 2014 Aug 14;512(7513):126–9. Available from: https://www.nature.com/articles/512126a

303. Ng LHX, Taeihagh A. How does fake news spread? Understanding pathways of disinformation spread through APIs. *Policy & Internet* [Internet]. 2021 Dec 1;13(4):560–85. Available from: https://doi.org/10.1002/poi3.268

304. Cardoso A, Leitão J, Teixeira C. Using the Jupyter notebook as a tool to support the teaching and learning processes in engineering courses. In: *The Challenges of the Digital Transformation in Education*. Auer ME, Tsiatsos T, editors. Cham: Springer International Publishing; 2019. p. 227–36.

305. Wang AY, Mittal A, Brooks C, Oney S. How data scientists use computational notebooks for real-time collaboration. *Proc ACM Hum-Comput Interact* [Internet]. 2019;3(CSCW). Available from: https://doi.org/10.1145/3359141

306. Zolkifli NN, Ngah A, Deraman A. Version control system: A review. *Procedia Comput Sci* [Internet]. 2018;135:408–15. Available from: https://www.sciencedirect.com/science/article/pii/S1877050918314819

307. Rule A, Birmingham A, Zuniga C, Altintas I, Huang S-C, Knight R, et al. Ten simple rules for reproducible research in Jupyter notebooks. *arXiv Prepr* arXiv181008055. 2018;

308. Saltelli A, Andres TH, Homma T. Sensitivity analysis of model output: An investigation of new techniques. *Comput Stat Data Anal.* 1993;15(2):211–38.
309. Razavi S, Jakeman A, Saltelli A, Prieur C, Iooss B, Borgonovo E, et al. The future of sensitivity analysis: An essential discipline for systems modeling and policy support. *Environ Model Softw.* 2021;137:104954.

Index

Printed in the United States
by Baker & Taylor Publisher Services